# MACHINE LEARNING APPLICATIONS FOR DATA ANALYSIS IN HEALTHCARE SYSTEMS

# MACHINE LEARNING APPLICATIONS FOR DATA ANALYSIS IN HEALTHCARE SYSTEMS

**Sudeshna Chakraborty, PhD**
**Jyotsna Singh, PhD**
**Praveen Kumar Shukla, PhD**
**Prasenjit Chatterjee, PhD**
*Editors*

AAP | APPLE ACADEMIC PRESS

First edition published 2026

**Apple Academic Press Inc.**
1265 Goldenrod Circle, NE,
Palm Bay, FL 32905 USA

760 Laurentian Drive, Unit 19,
Burlington, ON L7N 0A4, Canada

**CRC Press**
2385 NW Executive Center Drive,
Suite 320, Boca Raton FL 33431

4 Park Square, Milton Park,
Abingdon, Oxon, OX14 4RN, UK



*Apple Academic Press exclusively co-publishes with CRC Press, an imprint of Taylor & Francis Group, LLC*



**Library and Archives Canada Cataloguing in Publication**

Title: Machine learning applications for data analysis in healthcare systems / Sudeshna Chakraborty, PhD, Jyotsna Singh, PhD, Praveen Kumar Shukla, PhD, Prasenjit Chatterjee, PhD, editors.
Names: Chakraborty, Sudeshna, editor. | Singh, Jyotsna, editor. | Shukla, Praveen Kumar, editor. | Chatterjee, Prasenjit, 1982- editor
Description: First edition. | Includes bibliographical references and index.
Identifiers: Canadiana (print) 20250222957 | Canadiana (ebook) 20250222973 | ISBN 9781779643186 (hardcover) | ISBN 9781779643193 (ebook)
Subjects: LCSH: Artificial intelligence—Medical applications. | LCSH: Machine learning. | LCSH: Medical informatics.
Classification: LCC R859.7.A78 M33 2026 | DDC 610.285—dc23

**Library of Congress Cataloging-in-Publication Data**

CIP data on file with US Library of Congress

ISBN: 978-1-77964-318-6 (hbk)
ISBN: 978-1-77964-319-3 (ebk)
DOI: 10.1201/9781779643193

# About the Editors

**Sudeshna Chakraborty, PhD**
*Professor, Department of Computer Science and Engineering, Galgotias University, Greater Noida, Uttar Pradesh, India*

Dr. Sudeshna Chakraborty is currently Professor and Research Group Head of data analytics and deep learning at the Department of Computer Science Engineering at Galgotias University, Greater Noida, Uttar Pradesh, India. She has over 20 years of rich academic and industry experience. She has acquired several awards as a distinguished professor, including a research excellence award, Engineers Award by the Institute of Engineers of India, amongst others. She has also been a keynote speaker, an organizing member of international conferences, a member of review committees, session chair with the Institute of Engineers, speaker at InSc- and AICTE Training and Learning (ATAL) Academy-sponsored and other faculty development programs, etc. She has filed eight patents in the field of robotic, solar energy, and sensors. She has chaired an IEEE conference in Paris, including *International Conference on Advances in Computing and Communication Engineering (ICACCE) , International Conference on Innovative Computing and Communication (ICICC), International Conference on Cyber Security and Artificial Intelligence (ICCSAI), and International Conference on Structures and Architecture (ICSA2025) (ICS2A)*. Dr. Chakraborty has been instrumental in various industrial interfacing for both academics and researchers at her previous assignments at various organizations (Sharda University, ManavRachna University, Mumbai University, Lingaya's, Institute of Chartered Financial Analysts of India [ICFAI], and others). She has over 150 publications in Scopus- and SCI-indexed high-impact journals and international conferences. She has published 20 patents, five of which have been granted. She is guiding five PhD scholars at various universities, two of whom have successfully completed their PhDs, along with several post- and undergraduate students. At the same time, she has earned prestigious accreditations from

NAAC, NBA, QAA, WASC, UGC, IAU, IET, and others. She holds MTech and PhD degrees in Computer Science and Engineering, and her area of expertise is neural network and semantic web engineering.

**Jyotsna Singh, PhD**
*Director (I/c) SVKM's NMIMS Chandigarh and Associate Dean, STME, NMIMS Chandigarh, India*

Dr. Jyotsna Singh is a BE, MTech, and PhD holder with over 22 years of experience in the education industry. She has held several senior academic and administrative positions such as Director, Deputy Dean Students, and more, across esteemed institutions including NIT Kurukshetra, NorthCap University, Amity University, Lloyd Group, IILM, among others.

She is currently serving as the Director (I/C), NMIMS Chandigarh, and Associate Dean, School of Engineering and Technology, NMIMS Chandigarh Campus.

Dr. Singh specializes in Computer Science and Engineering and holds professional certifications in Data Science, Python, Machine Learning, Artificial Intelligence, Computational Thinking, and Strategic Mindset from reputed institutions such as Wipro, IBM, Deeplearning.AI, University of Pennsylvania, and University of Michigan. She also completed a certification in High Impact Teaching Skills organized by Wipro and attested by Dale Carnegie & Associates.

In her illustrious career, she has led various workshops, undertaken government-funded projects, and launched several university-level initiatives. She has presented and published high-quality research in reputed journals and conferences and is proficient in programming languages such as C, C++, Python, among others.

She has also filed over five patents in the areas of AI, IoT, and smart systems.

An accomplished author, she has several departmental books to her credit, including Lab Manual on *Software Project Management, FOCP, and Data Structures*. Her contributions to the academic world were recognized when she received the "Torchbearer of Education" award in 2020 from Coding Ninjas.

**Praveen Kumar Shukla, PhD**
*Associate Professor, Department of IoT & Intelligent Systems, Manipal University Jaipur, Jaipur, Rajasthan, India*

Dr. Praveen Kumar Shukla graduated in 2012 from Lakshmi Narain College of Technology & Science, Bhopal. He earned his Master of Technology in Control and Automation Engineering from VIT, Vellore, in 2015 and completed his PhD in Brain-Computer Interfacing in 2021 from the National Institute of Technology (NIT), Raipur, India. In 2024, he completed his Postdoctoral Fellowship at the Indian Institute of Technology (IIT), Gandhinagar. With over 8 years of academic and research experience, Dr. Shukla currently serves as an Associate Professor (Selection Grade) in the Department of IoT & IS Engineering at Manipal University Jaipur. He has been awarded a research grant of ₹41 lakh from the Department of Science and Technology (DST) under the ASEAN-India Collaborative R&D Scheme. Dr. Shukla has an impressive research portfolio with over 45 publications in reputed journals, conferences, and book chapters spanning areas such as Brain-Computer Interfacing, Medical Image Processing, Data Mining, Machine Learning, and Deep Learning. Additionally, he has published six patents, three of which are granted, and has edited a book. He is actively involved in mentoring students, currently supervising two PhD scholars, three M.Tech students, and 15 B.Tech students. To date, he has successfully supervised 20 B.Tech and M.Tech candidates. Dr. Shukla also contributes as a reviewer for esteemed journals, including the *IEEE Journal of Biomedical and Health Informatics* and *Disability and Rehabilitation: Assistive Technology.*

**Prasenjit Chatterjee, PhD**
*Dean (Research and Consultancy), MCKV Institute of Engineering, West Bengal, India*

Dr. Prasenjit Chatterjee currently serves as Professor of Mechanical Engineering and Dean (Research and Consultancy) at MCKV Institute of Engineering (NAAC Accredited A Grade Autonomous Institute), Howrah, West Bengal, holds a PhD in Engineering from Jadavpur University and has completed postdoctoral research from Brazil and Norway. With over 8800 citations and 147 research papers in reputed international journals and conferences, he has authored and edited more than 54 books on intelligent decision-making, supply chain management, optimization, sustainability, and risk analysis. He has consecutively been recognized by Elsevier-Scopus and Stanford University's global top 2% scientists list in "Operations Research," Dr. Chatterjee has developed several widely cited MCDM methods like MARCOS, RAFSI, SRP, and PRASIAS. He has served as Guest Editor for various SCIE and Scopus indexed journals and leads multiple international book series with Springer, CRC Press, Scrivener-Wiley, and Apple Academic Press. He has received numerous awards including Best Track Paper Award at IEOM 2018 (Paris), Outstanding Reviewer Award from Elsevier, and a University Gold Medal. As a dedicated research mentor, he actively guides PhD scholars as a co-supervisor at institutions like NIT Patna and Aliah University, focusing on advanced decision-making techniques and sustainability research.

# Contents

# Contributors

**Shivam Agarwal**
Department of Computer Science & Engineering, Meghnad Saha Institute of Technology, Kolkata, India

**Toshini Agrawal**
School of Computing Science Engineering and Artificial Intelligence, VIT Bhopal University, Madhya Pradesh, India

**Sandhya Avasthi**
Department of Computer Science and Engineering, ABES Engineering College, Ghaziabad, Uttar Pradesh, India

**Rohit Bansal**
Department of Management, Vaish Engineering College, Rohtak, Haryana, India

**Prachi Bhatt**
School of Computing Science Engineering and Artificial Intelligence, VIT Bhopal University, Madhya Pradesh, India

**Jigyasa Bisht**
School of Computing Science Engineering and Artificial Intelligence, VIT Bhopal University, Madhya Pradesh, India

**Sudeshna Chakraborty**
Department of Computer Science and Engineering, Galgotias University, Greater Noida, Uttar Pradesh, India

**Prateek Chaurasia**
Department of Information Technology, Ajay Kumar Garg Engineering College, Ghaziabad, Uttar Pradesh, India

**Souvik Das**
Indian Institute of Technology, Kharagpur, West Bengal, India

**Ajitesh Gautam**
Department of Computer Science & Engineering, Galgotias College of Engineering and Technology, Greater Noida, Uttar Pradesh, India

**Ashish Garg**
Indian Institute of Technology, Kharagpur, West Bengal, India

**Pallavi Goel**
Department of Computer Science & Engineering, Galgotias College of Engineering and Technology, Greater Noida, Uttar Pradesh, India

**Agam Das Goswami**
School of Electronics Engineering, VIT-AP University, Andhra Pradesh, India

**Chabi Gupta**
Amity College of Commerce and Finance Amity University, Noida, India

**Suvam Gupta**
Department of Computer Science & Engineering, Meghnad Saha Institute of Technology, Kolkata, India

**Pallavi Jain**
Department of Computer Science and Engineering, Galgotias University, Greater Noida, Uttar Pradesh, India

**Keshav Kaushik**
Center for Cyber Security and Cryptology, Sharda School of Computer Science & Engineering, Sharda University, Greater Noida, India

**Ganesh Kumar M**
School of Electronics Engineering, VIT-AP University, Andhra Pradesh, India

**Anil Kumar**
Galgotias University, Greater Noida, Uttar Pradesh, India

**Amardeep Kumar**
Indian Institute of Technology, Kharagpur, West Bengal, India

**Vimal Kumar**
Department of Computer Science and Engineering, Madan Mohan Malaviya University of Technology, Gorakhpur, Uttar Pradesh, India

**Virendra Kushwah**
School of Computing Science Engineering and Artificial Intelligence, VIT Bhopal University, Madhya Pradesh, India

**J. Maiti**
Indian Institute of Technology, Kharagpur, West Bengal, India

**Pawan Kumar Mall**
Department of Computer Science and Engineering, Madan Mohan Malaviya University of Technology, Gorakhpur, Uttar Pradesh, India

**Subhash Mondal**
Department of Computer Science & Engineering (AI & ML), Dayananda Sagar University, Bangalore, India

**Vipul Narayan**
Department of Computer Science and Engineering, Madan Mohan Malaviya University of Technology, Gorakhpur, Uttar Pradesh, India

**M Niranjanamurthy**
Department of Computer Applications, M S Ramaiah Institute of Technology, Bangalore, Karnataka, India

**Fincy Pallissery**
School of Law, Christ University Delhi NCR Campus, Delhi, India

**Ganapati Panda**
Department of Electronics and Tele Communication, C V Raman Global University, Bhubaneswar, Odisha, India

**Kriti Parashar**
School of Law, Christ University Delhi NCR Campus, Delhi, India

**Adyasha Rath**
Department of Computer Science and Engineering, C. V. Raman Global University, Bhubaneswar, Odisha, India

**Shweta Roy**
Department of Computer Science and Engineering, ABES Engineering College, Ghaziabad, Uttar Pradesh, India

**Diganta Sengupta**
Department of Computer Science & Engineering, Heritage Institute of Technology, Kolkata, India

**Anupama Sharma**
Department of Information Technology, Ajay Kumar Garg Engineering College, Ghaziabad, Uttar Pradesh, India

**Praveen Kumar Shukla**
Associate Professor, Department of IOT & Intelligent Systems, Manipal University Jaipur, Jaipur, Rajasthan, India

**Pallavi Singh**
Department of Biotechnology, Graphic Era (Deemed to Be University), Dehradun, Uttarakhand, India

**Shantanu Singh**
Department of Computer Science & Engineering, Galgotias College of Engineering and Technology, Greater Noida, Uttar Pradesh, India

**Brajesh Kumar Singh**
Department of Electronics and Communication Engineering, Galgotias college of Engineering and Technology, Greater Noida, Uttar Pradesh, India

**Ram Singh**
School of Banking, Financial Services & Insurance Management, Symbiosis University of Applied Sciences, Indore, India

**Shivangi Singh**
School of Computing Science Engineering and Artificial Intelligence, VIT Bhopal University, Madhya Pradesh, India

**Somya Sinha**
Department of Biotechnology, Graphic Era (Deemed to Be University), Dehradun, Uttarakhand, India

**Akash Srivastava**
Department of Biotechnology, Graphic Era (Deemed to Be University), Dehradun, Uttarakhand, India

**Prashant Srivastava**
Department of Information Technology, Ajay Kumar Garg Engineering College, Ghaziabad, Uttar Pradesh, India

**Swapnita Srivastava**
Department of Computer Science and Engineering, Madan Mohan Malaviya University of Technology, Gorakhpur, Uttar Pradesh, India

**Anjey Mani Tripathi**
Department of Computer Science and Engineering, Galgotias University, Greater Noida, Uttar Pradesh, India

**Nivedita Upadhyay**
Department of Biotechnology, Graphic Era (Deemed to Be University), Dehradun, Uttarakhand

**Ashish Yadav**
Department of Computer Science & Engineering, Galgotias College of Engineering and Technology, Greater Noida, Uttar Pradesh, India

# Abbreviations

| | |
|---|---|
| AB | adaptive gradient boost |
| ACC | accuracy |
| AGm | adjusted geometric mean |
| AI | artificial intelligence |
| ANN | artificial neural network |
| AUC | area under the curve |
| AUROC | area under the receiver operating characteristic curve |
| AV | atrioventricular |
| BLPOC | band-limited phase only correlation |
| BNB | Bernoulli naive Bayes |
| BRF | balanced random forest |
| CAD | coronary artery disease |
| CADD | computer-aided drug design |
| CBT | cognitive behavioral technique |
| CHD | coronary heart disease |
| CNN | convolution neural network |
| CoV | coronavirus |
| CRP | C-reactive protein |
| CT | computed tomography |
| CV | cross-validation |
| CVD | cardiovascular disease |
| CWT | continuous wavelet transform |
| DAS28 | Disease Activity Score on 28 joints |
| DL | deep learning |
| DMARD | disease-modifying antirheumatic drug |
| DT | decision tree classifier |
| DWT | discrete wavelet transform |
| ECC | edge-cognitive-computing |
| ECG | electrocardiogram |
| EER | energy efficiency ratio |
| EHR | electronic health records |
| ESR | erythrocyte sedimentation rate |
| FCM | fuzzy c-mean segmentation |
| FDA | Food and Drug Association |
| FFT | fast Fourier transform |
| FKP | finger knuckle print |

| | |
|---|---|
| FN | false negative |
| FP | false positive |
| GAN | generative adversarial network |
| GB | gradient boosting |
| GNB | Gaussian naive Bayes |
| GORP | gradient ordinal relation pattern |
| HCPs | healthcare professionals |
| HD | heart disease |
| HF | heart failure |
| HGB | histogram-based gradient boosting |
| HTS | high-throughput screening |
| ICA | independent component analysis |
| ICG | impedance cardiography |
| ICU | intensive care units |
| IDF | inverse document frequency |
| IoT | Internet of Things |
| JSN | joint space narrowing |
| KNN | K-nearest neighbor |
| KOA | knee osteoarthritis |
| LASSO | least absolute shrinkage and selection operator |
| LBDD | ligand-based drug designing |
| LDA | linear discriminant analysis |
| LGIC | local global information combination |
| LR | logistic regression |
| LSTM | long short-term memory |
| MCG | magnetocardiography |
| MFCC | Mel-frequency cepstral coefficients |
| MFRT | modified finite radon transform |
| ML | machine learning |
| MR | magnetic resonance |
| MTX | methotrexate |
| NB | naïve Bayes |
| NER | named entity recognition |
| NLP | natural language processing |
| NLTK | Natural Language Toolkit |
| OA | osteoarthritis |
| OCLPP | orthogonal complex locality preservation projections |
| ODEs | ordinary differential equations |
| OLDA | orthogonal linear discriminant analysis |
| PCA | principal component analysis |
| PDEs | partial differential equations |
| PE | precision |

| | |
|---|---|
| PLI | power line interface |
| QSAR | quantitative structure–activity relationship |
| RA | rheumatoid arthritis |
| RANSAC | RANdom SAmple Consensus |
| RCC | renal cell carcinoma |
| RE | recall |
| RF | random forest |
| RFID | radio frequency identification |
| RMFRD | real-world masked face recognition datasets |
| RNN | recurrent neural network |
| ROC–AUC | receiver operating characteristic–area under the curve |
| SARS-CoV-2 | severe acute respiratory syndrome coronavirus 2 |
| SBDD | structure-based drug designing |
| SBVS | structure-based virtual screening |
| SGORP | star gradient ordinal relation pattern |
| SMOTE | synthetic minority oversampling technique |
| STFT | short-time Fourier transform |
| SURF | speeded-up robust features |
| SVC | support vector classifier |
| SVM | support vector machine |
| TEO | Teager energy operator |
| TF | term frequency |
| TN | true negative |
| TP | true positive |
| VEGF | vascular endothelial growth factor |
| VHA | virtual health assistants |
| VoIP | Voice over Internet Protocol |
| VR | virtual reality |
| WLE | weighted linear embedding |
| XGB | extreme gradient boosting |
| XGBoost | extreme gradient boosting |

# Acknowledgment

The editors would like to acknowledge the help of all the people involved in this project and, more specifically, the authors and reviewers that took part in the review process. Without their support, this book would not have become a reality.

First, the editors would like to thank each one of the authors for their contributions. Our sincere gratitude goes to the chapter authors who contributed their time and expertise to this book.

Second, the editors wish to acknowledge the valuable contributions of the reviewers regarding the improvement of quality, coherence, and content presentation of chapters.

Last but not the least, the editors are most grateful to the whole Apple Academic Press team, particularly to Ashish, Sandra, and Rakesh for their great cooperation, motivation, and support.

***The Editors***

# Preface

Welcome to *Machine Learning Applications for Data Analysis in Healthcare Systems*. This book is a comprehensive exploration of the powerful intersection between machine learning and healthcare. It investigates the ever-changing impact of machine learning techniques on data analysis in healthcare, transforming the way we approach medical challenges, improve patient outcomes, and enhance healthcare systems.

The healthcare industry generates an enormous amount of data, from electronic health records and medical imaging to genomic sequencing and wearable devices. However, the true value of this data lies not in its sheer volume but in the insights it can provide. Machine learning algorithms offer us the means to unlock the hidden patterns and knowledge within this data, enabling us to make informed decisions, identify high-risk patients, and personalize interventions for better healthcare outcomes.

This book aims to bridge the gap between theory and practice, providing a broad understanding of machine learning techniques specifically tailored for healthcare applications. Whether you are a healthcare professional, data scientist, researcher, or simply curious about the potential of technology in healthcare, this book will serve as an invaluable resource.

This book emphasizes on the practical implementation of machine learning techniques, supported by real-world case studies and examples. It also addresses the ethical considerations and challenges associated with implementing machine learning in healthcare, ensuring that responsible and ethical practices are at the forefront of the discussions.

The book *Machine Learning Applications for Data Analysis in Healthcare Systems* aims to provide readers with the knowledge and tools required to understand and apply machine learning in the context of healthcare. Through a focus on key principles, real-world challenges, and practical case studies, this book empowers readers to use machine learning techniques effectively and contribute meaningfully to the improvement of healthcare systems for patients and society.

Let the exploration of machine learning in healthcare begin!

This book contains 13 chapters on different aspects of machine learning applications for data analysis in healthcare system.

Chapter 1 of this book investigates the application of machine learning techniques for classifying in-hospital mortality among heart failure patients admitted to the intensive care unit (ICU), using limited historical data in low-resource environments. By employing advanced algorithms and feature engineering, it aims to develop a reliable and accessible predictive model. This research focuses on addressing the challenges posed by scarce data availability, aiming to enhance risk assessment and inform clinical decision-making, thereby improving patient outcomes and resource allocation in ICU environments.

In the chapter 2 this book explores machine learning and its potential in outbreak prediction. By using the power of advanced algorithms and data-driven approaches, it aims to forecast and identify potential disease outbreaks more accurately and rapidly. It explores various methodologies, data sources, and predictive models, highlighting the significance of early detection and proactive response strategies to mitigate the devastating impact of infectious diseases on public health and global well-being.

Chapter 3 explores the development of a mobile application designed to analyze and predict depression in individuals. By harnessing the power of data-driven approaches and machine learning algorithms, this app aims to provide an accessible and user-friendly tool for early detection and monitoring of depressive symptoms. It involves the integration of various data sources, including self-reported information and smartphone sensor data, to enhance the accuracy and reliability of depression prediction. The ultimate goal is to empower individuals and healthcare professionals with a valuable resource for proactive mental health management.

Chapter 4 presents a comprehensive review of machine learning-based analytics for the early detection of premature rheumatoid arthritis (RA) and osteoarthritis (OA) in clinical practices. By examining existing literature and studies, we explore the potential of advanced algorithms in identifying key biomarkers and patterns indicative of these debilitating joint diseases. The review emphasizes the importance of early diagnosis for effective management and intervention strategies. This chapter aims to provide valuable insights and guidelines for integrating machine learning approaches into routine clinical workflows, ultimately improving patient outcomes and quality of life.

Chapter 5 focuses on the performance evaluation of heart disease classification using deep learning and machine learning methods. By analysing and comparing the efficacy of various algorithms and models, we aim to identify the most accurate and reliable techniques for diagnosing heart diseases. The evaluation encompasses diverse datasets, feature engineering strategies, and performance metrics, shedding light on the strengths and limitations of different approaches. The findings presented in this chapter provide valuable insights for researchers and practitioners in the field, aiming to enhance the early detection and management of heart diseases using advanced analytical methods.

Chapter 6 explores the application of artificial intelligence (AI) in the healthcare sector, highlighting its benefits and challenges. By examining real-world use cases, we delve into how AI technologies, including machine learning and deep learning, are revolutionizing healthcare delivery, diagnostics, treatment, and patient outcomes. It discusses the potential advantages of AI, such as improved accuracy, efficiency, and personalized care. However, we also address the challenges surrounding data privacy, ethics, algorithm interpretability, and integration into existing healthcare systems. This chapter aims to provide a comprehensive overview of the opportunities and obstacles in leveraging AI for transformative healthcare advancements.

Chapter 7 examines into the design and development of anti-cancerous drug molecules using structure and ligand-based drug designing computational approaches. By using advanced computational techniques, we explore the vast chemical space to identify potential drug candidates with enhanced efficacy and reduced side effects. The chapter highlights the significance of structure-based and ligand-based approaches in rational drug design, focusing on molecular modelling, virtual screening, and quantitative structure-activity relationship (QSAR) analysis. This research aims to accelerate the discovery and optimization of novel anti-cancer drugs, ultimately contributing to the fight against cancer and improving patient outcomes.

Chapter 8 provides a comprehensive review of smart healthcare systems, exploring their design, implementation, and impact on healthcare delivery. By examining the integration of advanced technologies, such as Internet of Things (IoT), wearable devices, and data analytics, we delve into the transformative potential of smart healthcare. The review encompasses various aspects, including remote patient monitoring, telemedicine, electronic health records, and healthcare management systems. Furthermore, we discuss the benefits, challenges, and future prospects of smart healthcare systems, aiming to provide valuable insights for researchers, policymakers, and healthcare professionals interested in harnessing technology for improved patient care and outcomes.

Chapter 9 explores the transformative potential of technology-enabled smart healthcare in shaping a Smart Society 5.0. By integrating cutting-edge technologies such as Internet of Things (IoT), Artificial Intelligence (AI), and Big Data analytics, it examines how smart healthcare systems can enhance patient care, optimize resource utilization, and empower individuals to take control of their health. The chapter discusses the various applications of technology-enabled healthcare, including remote patient monitoring, personalized medicine, and predictive analytics. Additionally, it explores the societal implications, ethical considerations, and challenges associated with the implementation of Smart Society 5.0. This research aims to inspire innovative solutions and drive the evolution of healthcare towards a smarter and more sustainable future.

Chapter 10 introduces ELM-HC, a novel approach for heartbeat classification based on a machine-human interaction model. By combining the strengths of machine learning algorithms and human expertise, we explore how collaborative decision-making can enhance the accuracy and interpretability of heartbeat classification. The chapter discusses the methodology, model architecture, and data pre-processing techniques used in developing ELM-HC. Furthermore, we examine the potential applications of this approach in healthcare settings, such as arrhythmia diagnosis and cardiovascular risk assessment. This research aims to bridge the gap between automated machine learning models and human expertise, providing a valuable tool for improved cardiac health monitoring and patient care.

Chapter 11 proposes an integrated natural language processing and Bow-Tie-based risk assessment methods for analysing accident reports.

Chapter 12 presents contexts regarding emerging biometric recognition methodology for identified people using Finger knuckle print technology.

Chapter 13 deals with face mask detection using deep learning and image processing algorithms. By combining Resnet-50 by bilinear interpolation, bilateral filter and fuzzy C means segmentation.

We would like to express our heartfelt gratitude to our parents, whose unwavering support and encouragement have been instrumental in our academic pursuits. Their belief in us has been a constant source of motivation.

We would also like to acknowledge the invaluable contributions of our colleagues and friends, who have provided guidance, insightful discussions, and collaborative efforts throughout this endeavour. Their expertise and feedback have been pivotal in shaping the content of this book.

Lastly, we extend our gratitude to the readers, researchers, and healthcare professionals who will engage with this book. It is our sincere hope that the knowledge and insights shared within these pages will inspire further exploration, innovation, and advancements in the intersection of machine learning and healthcare data analysis.

Together, let us embrace the transformative power of machine learning and pave the way for a future where data-driven approaches revolutionize healthcare systems, leading to enhanced patient outcomes and well-being.

***—Editors***

# CHAPTER 1

# Classification of In-Hospital Mortality for Heart-Failure Patient Using Resource Constraint Dataset

DIGANTA SENGUPTA[1*], SUBHASH MONDAL[2], SUVAM GUPTA[3], and SHIVAM AGARWAL[3]

[1]*Department of Computer Science & Engineering, Heritage Institute of Technology, Kolkata, India*

[2]*Department of Computer Science & Engineering (AI & ML), Dayananda Sagar University, Bangalore, India*

[3]*Department of Computer Science & Engineering, Meghnad Saha Institute of Technology, Kolkata, India*

*Corresponding author

## ABSTRACT

The real challenge posed to in-hospital mortality study is from heterogeneous data from nonhomogeneous class of patients. This motivates the study, which uses traditional machine learning (ML) algorithms to analyze the heterogeneous data containing 48 features and classify in-hospital mortality. The retrospective cohort was focused on heart failure patients. Twelve ML algorithms have been deployed on a publicly available dataset for binary classification. The dataset used for the study was highly resource constrained and unbalanced; hence a novel data preprocessing unit has been designed. In addition to standard implementations, hyperparameter tuning and stacked implementations have also been done. Post cross-validation, the accuracy value observed was 95.45% for histogram-based Gradient

Machine Learning Applications for Data Analysis in Healthcare Systems: Sudeshna Chakraborty, Jyotsna Singh, Praveen Kumar Shukla, Prasenjit Chatterjee (Eds.)


DOI: 10.1201/9781779643193-1

Boost algorithm. Hyper-parameter-tuned model accuracy was observed at 96.55% for the Gradient Boost classifier. Accuracy for the stacked model was 96.43%, with an RoC-AuC score of 0.977, a recall value of 0.964, and an F1-Score of 0.977. The false negative share for the stacked model was at 0.56%. Since mortality analysis is the subject of study, we claim that the false negative value validates the robustness of the classification. The performance of all the ML algorithms has been compared with the existing literature.

## 1.1 INTRODUCTION

Heart failure is a complicated clinical ailment resulting in hindrance of ventricular filling or emission of blood to the systemic circulation due to structural or functional heart disorder. By definition, it is failure to meet the systematized demands of circulation. It implies that there is some anomaly present in the systolic or diastolic phases of the heart which is hampering the heart's performance and in turn it can lead to huge consequences later on. It has affected many people globally and the death rate associated with it is increasing day by day. It has an estimated ubiquity of around 26 million people globally and it contributes to an escalated healthcare cost worldwide. Heart failure is a common, expensive, and potentially lethal condition having a mortality rate of around 4–7% [1], and it is one of the major causes of both hospitalization and readmission. The heart-failure rate in adults and senior citizens is around 2–3% and 6–10%, respectively, and is bound to increase in the coming years. Also, the death rate is about 35% after the diagnosis of heart disease after one year.

Nowadays, many machine learning (ML) models and severity scores are being utilized in the medical domain but the majority of them give unsatisfactory results which will not be helpful in predicting mortality effectively [2]. Recently, many ML models and severity scores are being utilized in the medical domain but the majority of them give unsatisfactory results which will not be helpful in predicting mortality effectively. Above 80% results have been noted for the classification of disease by using ML methodologies [3].

Our one and only motive is to provide a model that would be able to correctly predict the chances of a person surviving this lethal condition which would be helpful in providing extra care to people having a high risk of mortality, with correct medical supervision and equipment's which we can only find in the intensive care units (ICU). Hence, a stacking ensemble model is proposed here, consisting of six base learners (random forest [RF], extreme gradient boosting [XGB], support vector machine [SVM], *K*-nearest neighbors [KNN], decision tree classifier [DT], adaptive gradient boost [AB], gradient boosting [GB], and histogram-based gradient boosting [HGB]) and light gradient boosting machine (LGBM) as metalearner which we have chosen from a pool of 12 ML classification algorithms such as AB, GB, LGBM, SVM, RF, logistic regression (LR), DT, XGB, HGB, Gaussian naive Bayes (GNB), Bernoulli naive Bayes (BNB), and KNN.

The various ML proposed model has less significant performances in majority of the cases related to in-hospital mortality prediction in ICU patients [4]. In this regard, we have observed many difficulties that all the medical datasets have two things in common and that is highly unbalanced and data with lots of missing values [5, 6]. Data extracting is a more challenging task to control the high volume of missing values within the dataset due to irregular sampling [7]. Data imbalance makes the training procedure more hectic and in turn deteriorates the performance of the model in prognosticating the outcome [8]. Also, it is stated by Anand et al. [9] that each and every algorithm has its own advantages and drawbacks. So before applying any algorithms, the data were processed thoroughly and all the missing values were imputed and the data were balanced and scaled so that the results do not get biased or degraded by any means [10, 11].

After preprocessing the data, we divided the data into train and test splits and trained all of our 12 ML algorithms mentioned above using our training data having *K*-fold cross-validation (CV) of value 10 and then tested all our models. After this, we finely tuned the models that were yielding better results. Thereafter, testing and evaluating many different combinations of tuned models for metalearner and base learners of our stacking ensemble model, we got the best results while using LGBM as metalearner and RF, XGB, SVM, KNN, DT, AB, GB, and HGB as base learners. Hence, declaring this as our proposed model which achieves an accuracy of 0.982, $F_1$ score 0.977, area under the receiver operating characteristic curve (AUROC) 0.977, recall 0.969, precision 0.984, Cohen kappa 0.955, standard deviation 0.64.

## 1.2 BACKGROUND ON ML ALGORITHMS

### *1.2.1 SUPPORT VECTOR MACHINE*

SVM is an algorithm which comes under the supervised category of ML algorithms which finds its place in solving both classification and regression tasks [12] but mostly preferred for classification tasks. This algorithm makes a clear line of distinction known as hyperplane between the data points that gets plotted according to the input data and it classifies data based on the distinction made by the corresponding hyperplane. This algorithm is fast to train unlike other algorithms based on trees. It can also be used on datasets with many rows and columns. It generally gives good results for the datasets where the distinction between data points can be made easily.

### *1.2.2 RANDOM FOREST*

This algorithm is present in the supervised category of ML algorithm which is predominantly used in both regression and classification type problems [13]. Multiple DTs are generated on the training data and then the most voted or most prevailing result is considered as the final result. Since RF uses multiple DTs to train the algorithm by selecting the data in a random manner, it ensures that every data contributes their meaningful insights toward the accurate model formation which gives efficient results. It also ensures that the model does not overfit by considering the majority of the votes.

### *1.2.3 DECISION TREE*

It is a tool that has applications spanning various different areas [14]. This algorithm is also a part of the supervised algorithm family of ML which deals with regression as well as classification problems. It is the most fundamental algorithm which forms the basis of many different ML algorithms like RF, gradient-boosted trees, XGB, and so on. It can handle both categorical and numerical data. Each node of this algorithm represents a decision that is to be taken, to go further down the tree and reach the leaf node that represents the result or the outcome of the classification task. It cleaves its node focusing on concepts like information gain, Gini index, and entropy. Its aim is to divide the whole feature space and infer meaningful information from it and create decision nodes that help in fabricating the overall model and hence prognosticating the outcome of the problem.

### *1.2.4 LOGISTIC REGRESSION*

LR is one of the robust members of the supervised ML algorithm family which is easy to implement [15]. It always finds its usage whenever there is a discussion for a binary classification task as it is based on a sigmoid function which only has two outcomes either zero or one. It is trouble-free to execute, explain, and very efficient to train. It makes no presumptions in the feature space about distributions of classes and it can be easily prolonged to multiple classes. We can also customize the default threshold value which is 0.5 to any number between 0 and 1 whenever itis required to meet the demands of our problem and achieve better performance.

### *1.2. GAUSSIAN NAIVE BAYES*

GNB is also a supervised ML algorithm whichis based on the principles of naive Bayes and Gaussian normal distribution [16]. It is generally used for classification tasks but it can be used for regression tasks as well if some changes are made in the algorithm. A presumption is often taken that the contiguous values analogous to each class are disseminated according to normal distribution while working with contiguous data. It can also give high accuracy in comparison to other algorithms in datasets having less data as it does not require high amounts of data to be trained on. It can be used on datasets with high dimensionality as its training time is very less and its unresponsiveness to outliers makes it a highly robust, efficient, and reliable algorithm.

### *1.2.6 BERNOULLI NAIVE BAYES*

BNB is another algorithm based on the principles of naive Bayes and uses Bernoulli distribution to predict the outcome [16]. As it is based on Bernoulli distribution, it finds its usage when the data are discontinuous.

In this algorithm, all features are given the same priority and as the features are not interrelated; they can independently take part in prognosticating the final outcome, no feature has fear of getting left behind as the model penalizes itself for doing so. It sets aside the values that are continuous and only accounts for discrete or discontinuous data. It is blazingly fast in comparison with other classification algorithms, as it treats each feature separately with its corresponding binary values and it explicitly penalizes the model for absence of any of its features which is crucial for speculating the outcome. Due to its rapid prediction capability, it can find its usage in making real-time predictions and it is unresponsive to irrelevant features.

### *1.2.7 K-NEAREST NEIGHBOR*

KNN is a kind of algorithm which is merely based on intuition that data belonging to a similar group would have a similar impact on new data points' outcome which belongs to the same group [17]. This algorithm belongs to the supervised category of ML algorithm and is majorly used in classification use cases. It does not have any parameter to train according to the dataset, instead, it stores all the data that it gets and utilizes all those stored at the time of classification. It is one of the simplest ML algorithms and requires less time to train as compared to tree-based algorithms. It can also be used for regression problem statements by making some changes in the underlying algorithm.

### *1.2.8 ADAPTIVE BOOSTING CLASSIFIER*

Adaptive boosting is the most basic boosting ensemble algorithm which also belongs to the supervised category of the ML algorithms and canbe used for both classification as well as regression use cases [18]. It used multiple similar kinds of models to prognosticate the final outcome. The models that are used in fabricating this model are also known as weak learners. Each weak learner that is appended to the model with updated values of weight, tries to diminish the error of the previous model as a whole and at the end due to this the overall performance of the model increases, hence it can predict the outcome more accurately. It has proven to give better results as compared to DTs as the aggregation makes it more robust and increases the overall performance of the model.

### 1.2.9 GRADIENT BOOST

This is one of the most robust algorithms in ML [19]. GB is an ML procedure which has its applications in both regression and classification tasks. It furnishes a prognostic model which is basically DTs in manifestation of an ensemble of weak prediction models. GB generally surpasses RFs because it minimizes the errors by constructing a new model each time on the flaws of the preceding models. A gradient-boosted tree is fabricated in a hierarchical manner as in other boosting procedures, but it globalizes other procedures by optimizing randomly differentiable loss functions. The bias error of the model can be diminished by using a GB algorithm and therefore it enhances the efficiency in predicting the outcome of our binary classification task and hence it increases the accuracy of our overall model.

### 1.2.10 LIGHT GRADIENT BOOSTING MACHINE

LGBM is another boosting algorithm which comes under the family of supervised ML, and it is widely used in the industry because of its better performance [20]. It is highly memory efficient and takes less time to train with respect to the majority of the ML algorithms. It can be used for classification as well as regression use cases. This algorithm is based on trees, it rapidly decreases the error of the overall model as it always tries to split its nodes leaf-wise fashion hence fabricating the model leaf-wise and continuing to grow the leaf with high loss in order to reduce it. This is the reason why it performs better than other algorithms which are based on trees performing level-wise splits. LGBM algorithm uses histograms to bucket out contiguous feature values into distinct bins which enhances the speed of the training procedure; it also results in minimized memory usage. As compared to XGB, it is adept in executing equally well with enormous datasets with a remarkable depletion in training time. All these features made this algorithm the best fit for our dataset.

### 1.2.11 HISTOGRAM-BASED GRADIENT BOOSTING

HGB was introduced as an alternative to the GB algorithm's huge time-consuming nature while training on the dataset [21]. GB slows down the training procedure when the data has numerous rows. So to overcome this drawback, HGB was introduced which was also fabricated using trees and it converted the contiguous input data into individual distinct discrete packets and stored them in a histogram-like structure which helps it in saving time during the formation of trees. This model building approach using histograms reduces the overall model training time by many folds. This ML ensemble also comes under supervised machine category and can be used for classification as well as regression tasks. This approximator has indigenous support for absent (NaNs) values.

### *1.2.12 EXTREME GRADIENT BOOSTING*

XGB is a supervised ML algorithm which can be used for both classification and regression use cases. It is built to perform tasks with high efficiency and is easy to use and supple [22]. Under the GB framework, it applies the ML algorithms. XGB solves several data science tasks in a brisk and precise manner because it provides boosting of trees parallelly (also known as GBDT, GBM). Having an enhanced performance and pace in tree-based ML algorithms, itis one of the popular GB ensemble methods. It smartly and efficiently manages the memory for large datasets exceeding RAM, supports parallel processing, having a diversity of regularizations which help in minimizing overfitting, has good cache optimization, supports auto tree pruning which implies that the DTs will not grow further after certain limits internally, has a built-in CV, can handle outliers and missing values up to some extent. All these features made XGBoost suitable for our dataset.

## 1.3 RELATED WORK

In Ref. [1], authors have proposed a nomogram that will be able to prognosticate the in-hospital mortality of heart-failure patients which are admitted in the ICU of the hospitals. In the data preprocessing part, they imputed all the missing values in their dataset using the mean of their respective feature column. The ML algorithms that they used in their study were (XBG) and Least Absolute Shrinkage and Selection Operator (LASSO) regression. They had also used the GWTG-HF risk model. At the end, they chose XGB as their final model as it was performing better than the rest of the models, the area under receiver operating characteristic curve was 0.8416.

In Ref. [23], a stacking ensemble model was proposed by authors who foretell the mortality of the patients admitted in the ICU. They used five different algorithms, that is, KNN, LR, linear discriminant analysis for dimensionality reduction, DT, and multilayered perceptron on a time-stamped dataset of 10,664 patients having 80 features. Six different subsets of data were created from the most important features column. Before undergoing any undersampling technique to balance the dataset, the data werethoroughly preprocessed by handling missing values, using normalization, and removal of any noise present in the dataset. Their model sees an uplift in the performance on implementing a stacking model where LR acts as a metalearner. They achieved AUROC 93.3%, precision 96.4%, accuracy 94.4%, recall 91.1%, and $F_1$ score 93.7% on using *k*-fold CV evaluation process. In Ref. [24], the authors proposed to build a model whose decision-making capability is much better as compared to others existing models so as to properly assign the beds in the ICU to thepatients in need. They trained their models using two algorithms, that is, linear discriminant analysis classifier and SVM. Before undergoing the training phase, chi-square distribution was used to refine the data and SMOTE (synthetic minority oversampling technique) was used to deal with the imbalanced nature of the dataset. A total of 12 features wereused for prognosticating the mortality and a *k*-fold CV value of five was used to evaluate the SVM model. Between their presented models in the chapter, SVM works better with a recall of 69%, accuracy 77%, and fall out (0.5).

In Ref. [25], they have proposed a model that is useful in prognosticating extubation failure. The ML algorithm that they used for fabricating their model was LGBM. They have also shown that this algorithm performs better than algorithms like LR, XGB, ANN, and SVM. In the data preprocessing part, they had removed the feature that had missing values greater than 40% and imputed the rest using the mean of their respective column. They also discarded the data that were repetitive using Pearson's correlation coefficient. At last, they trained LGBM on top 36 features, obtained from feature importance analysis of 68 features which was interpreted using SAPS, XGBoost, and LGBM. They achieved an accuracy (80.2%), sensitivity (74.77%), specificity (83.94%), and AUROC (81.98%). In Ref. [26], the authors had presented a model for in-hospital mortality prediction using a balanced random forest (BRF) for ICU patients. They had taken into account the complication that comes with unbalanced data. They derived their data from the MIMIC-III database and chose seven distinct time series data from it and then extracted six statistical features from each of them which contributed to 716 features and applied LR and BRF on them. The hyperparameters of this algorithm were tuned using adjusted geometric mean (AGm). The model that was created using the BRF algorithm was performing well than the LR model as it was able to classify the inferior (class that was lesser in number) class with increased results incomparison to LR or any other benchmarking model with recall 69.52%, AUC (73.68%), and AGm (75.58%).

## 1.4 PROPOSED METHODOLOGY

In this section, we have discussed about the dataset that we have considered to build up ML model, different data preprocessing techniques to stabilize and clean the dataset for better performance of the ML model and also proposed an ensemble stacking-based ML model to predict the HF patients' mortality in the ICU. The workflow diagram of the proposed model is presented in Figure 1.1.

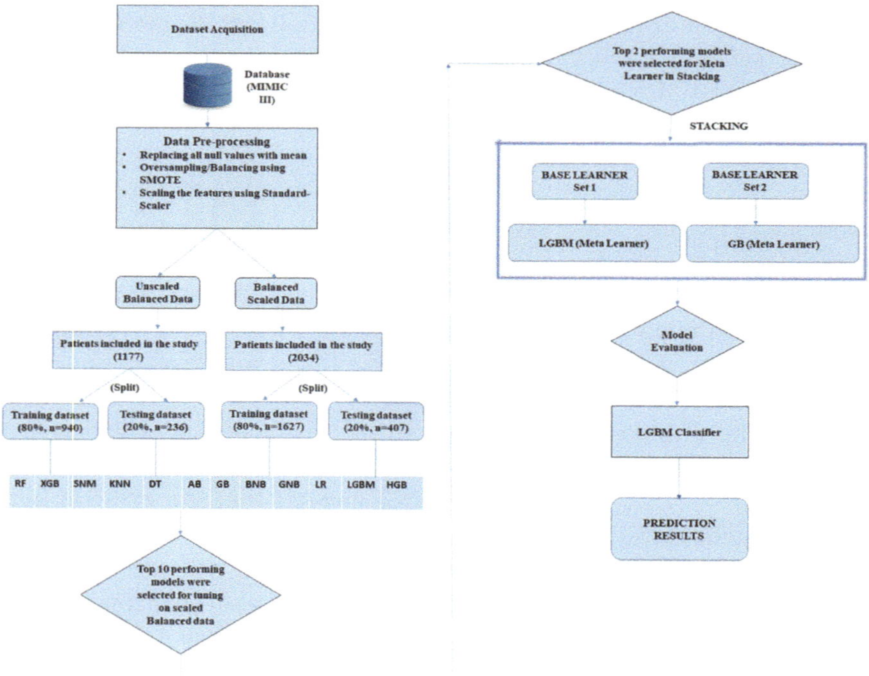


**FIGURE 1.1** Proposed workflow diagram.

### *1.4.1 DATASET ACQUISITION*

We analyzed the ICU mortality of heart-failure patients' dataset [27] and got to know that it had a total of 1177 instances of patients having 51 features in which 558 were male and 618 were females. The total number of death cases were 159 in which 80 were males and 79 were females. In general, the total alive rate was 86.5% and death rate was 13.5%. The "group" and "id" columns were discarded from our dataset as it had no use in predicting the outcome. Features details of the dataset with null values instances are presented in Table 1.1. Heatmap also representing the correlation of the features among all the features are in Figure 1.2.

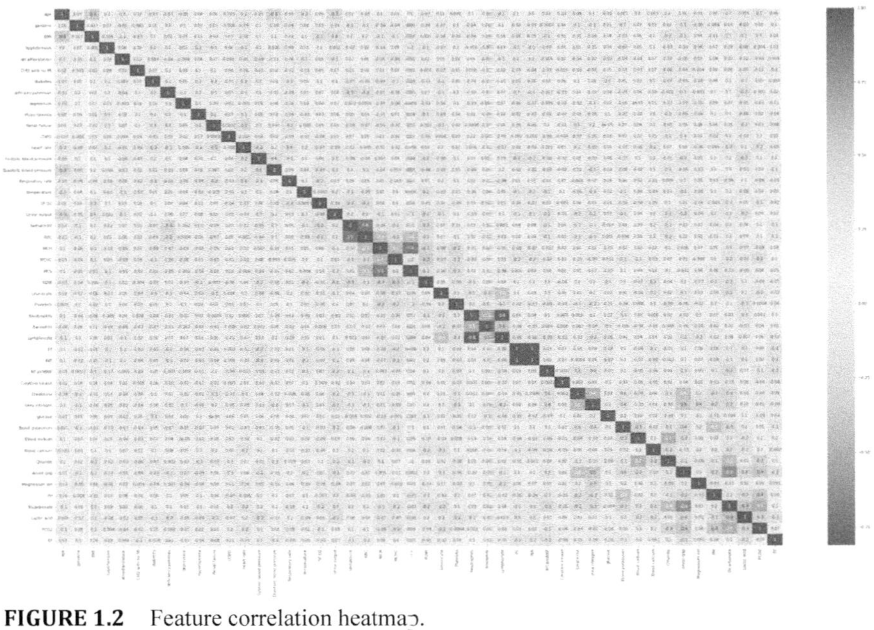

**FIGURE 1.2** Feature correlation heatmap.

### *1.4.2 DATA PREPROCESSING*

To clean the dataset, we performed some data preprocessing approaches before building a stable and high-performance ML model in terms of different matrices.

**TABLE 1.1** Values of All the Features in the Dataset

| Index | Features | Data Type | Non-null Values | Missing Values | Null Percentage |
|---|---|---|---|---|---|
| 1. | Group | int64 | 1177 | 0 | 0 |
| 2. | ID | int64 | 1177 | 0 | 0 |
| 3. | Outcome | float64 | 1176 | 1 | 0.08 |
| 4. | Age | int64 | 1177 | 0 | 0 |
| 5. | Gender | int64 | 1177 | 0 | 0 |
| 6. | BMI | float64 | 962 | 215 | 18.27 |
| 7. | Hypertensive | int64 | 1177 | 0 | 0 |
| 8. | Atrial fibrillation | int64 | 1177 | 0 | 0 |
| 9. | CHD with no MI | int64 | 1177 | 0 | 0 |
| 10. | Diabetes | int64 | 1177 | 0 | 0 |
| 11. | Deficiency anemia | int64 | 1177 | 0 | 0 |
| 12. | Depression | int64 | 1177 | 0 | 0 |
| 13. | Hyperlipemia | int64 | 1177 | 0 | 0 |
| 14. | Renal failure | int64 | 1177 | 0 | 0 |
| 15. | COPD | int64 | 1177 | 0 | 0 |
| 16. | heart rate | float64 | 1164 | 13 | 1.10 |
| 17. | Systolic blood pressure | float64 | 1161 | 16 | 1.36 |
| 18. | Diastolic blood pressure | float64 | 1161 | 16 | 1.36 |
| 19. | Respiratory rate | float64 | 1164 | 13 | 1.10 |
| 20. | Temperature | float64 | 1158 | 19 | 1.61 |
| 21. | $SPO_2$ | float64 | 1164 | 13 | 1.10 |
| 22. | Urine output | float64 | 1141 | 36 | 3.06 |
| 23. | Hematocrit | float64 | 1177 | 0 | 0 |
| 24. | RBC | float64 | 1177 | 0 | 0 |
| 25. | MCH | float64 | 1177 | 0 | 0 |
| 26. | MCHC | float64 | 1177 | 0 | 0 |
| 27. | MCV | float64 | 1177 | 0 | 0 |
| 28. | RDW | float64 | 1177 | 0 | 0 |
| 29. | Leukocyte | float64 | 1177 | 0 | 0 |
| 30. | Platelets | float64 | 1177 | 0 | 0 |
| 31. | Neutrophils | float64 | 1033 | 144 | 12.23 |
| 32. | Basophils | float64 | 918 | 259 | 22.01 |

**TABLE 1.1** *(Continued)*

| Index | Features | Data Type | Non-null Values | Missing Values | Null Percentage |
|---|---|---|---|---|---|
| 33. | Lymphocyte | float64 | 1032 | 145 | 12.32 |
| 34. | PT | float64 | 1157 | 20 | 1.70 |
| 35. | INR | float64 | 1157 | 20 | 1.70 |
| 36. | NT-proBNP | float64 | 1177 | 0 | 0 |
| 37. | Creatine kinase | float64 | 1012 | 165 | 14.02 |
| 38. | Creatinine | float64 | 1177 | 0 | 0 |
| 39. | Urea nitrogen | float64 | 1177 | 0 | 0 |
| 40. | Glucose | float64 | 1159 | 18 | 1.53 |
| 41. | Blood potassium | float64 | 1177 | 0 | 0 |
| 42. | Blood sodium | float64 | 1177 | 0 | 0 |
| 43. | Blood calcium | float64 | 1176 | 1 | 0.08 |
| 44. | Chloride | float64 | 1177 | 0 | 0 |
| 45. | Anion gap | float64 | 1177 | 0 | 0 |
| 46. | Magnesium ion | float64 | 1177 | 0 | 0 |
| 47. | pH | float64 | 885 | 292 | 24.81 |
| 48. | Bicarbonate | float64 | 1177 | 0 | 0 |
| 49. | Lactic acid | float64 | 948 | 229 | 19.46 |
| 50. | $PCO_2$ | float64 | 883 | 294 | 24.98 |
| 51. | EF | int64 | 1177 | 0 | 0 |

### *1.4.2.1 HANDLING MISSING VALUES*

We saw in Table 1.1 that a lot of features had null values. The number of missing values in any of the columns did not exceed 25%. So, they were to be filled in order to avoid any decay in accuracy or any biased behavior. So, keeping this in mind, the data were divided into two parts based on its outcome “alive” or “dead” which were denoted by “0” and “1,” respectively, in the outcome or the target column. Then all the missing feature values of each separated outcome data were handled using Simple Imputer having its strategy attributefixed to “mean” which means “mean” of their respective feature columns. After handling all the missing values, the datasets were concatenated and finally the outcome or target column was separated from the dataset.

### *1.4.2.2 BALANCING DATA*

As we can perceive from the data presented in Figure 1.3, we can see that our data are highly unbalanced with a ratio of approximately 1:6 for death to alive respectively. So, this means that our data are highly imbalanced and there is a need for balancing the data to avoid any chances of model biasing toward the majority class and hence deteriorating the overall performance of the model. This study of the dataset implies that we need to balance our data to achieve prime performance from our model. As we know, there are two ways to perform data balancing procedures one of that is undersampling, which randomly chooses the data of majority class and deletes it until both our data are equal and second technique is oversampling, which increases the data of minority class by making copies of the data present in it and hence increasing it until both data becomes equal. We rejected the undersampling technique because while deleting the data values, it might delete some data which carry meaningful insights toward the result formation of the model so we chose an oversampling technique and balanced our data using SMOTE. After applying SMOTE, the data distribution is presented in Table 1.2.

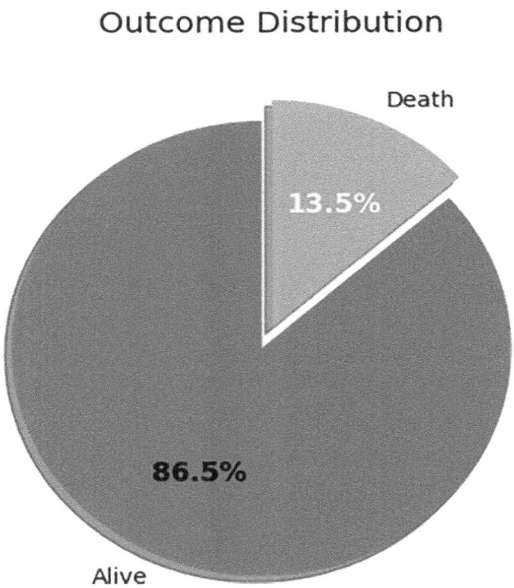


**FIGURE 1.3** Outcome distribution.

**TABLE 1.2** Data Distribution Before and After Balancing

| | 0 (Alive) | 1 (Dead) |
|---|---|---|
| Before balancing | 1017 | 159 |
| After balancing | 1017 | 1017 |

Now we had two datasets, one unbalanced and the other balanced. Then we created a training and testing split for both of our datasets with 80% and 20% of the data respectively using *sklearn* [28] train-test split. After splitting the data distribution is shown in Table 1.3.

**TABLE 1.3** Data Distribution after Forming Training and Testing Split

| | Training Split | Testing Split |
|---|---|---|
| Balanced | 1627 | 407 |
| Unbalanced | 940 | 236 |

#### *1.4.2.3 SCALING THE DATA*

As we can see in our dataset that the data vary very much and the difference between the lowest and the highest values is very huge. So, this could increase the overall model training time or may cause the model to get biased toward the data or the feature that has a huge positive or negative valuewhich in turn would make our model more prone to overfitting, thereby deteriorating the overall model's performance which we do not want to happen. Knowing the consequences that come with the unscaled data we thought of scaling our dataset and then training our model on the scaled data for making our model more robust and reducing the overall training time. Hence, we decided to scale our training and testing data of both unbalanced and balanced dataset using Standard Scalar. After applying Standard Scalar, the feature values of our dataset came in close proximity with each other and started ranging between −1 and 1.

### *1.4.3 MODEL TRAINING*

After all the preprocessing steps were done, we trained AB, GB, LGBM, XGB, LR, SVM, HGB, GNB, BNB, KNN, RF, and DT algorithms using the training data and then tested our models on the testing split of both the balanced and unbalanced dataset and then evaluated the performance of each model using seven different performance evaluation matrices such as receiver operating characteristic–area under the curve (ROC–AUC) scores, Cohen kappa, $F_1$ score, precision, recall, $K$-fold mean accuracy, and standard deviation and stored the output of them in Tables 1.4 and 1.5 for any future reference. While training and evaluating the model on the train and test dataset, we used the $K$-fold CV of 10 folds. XGB performed best accuracy of 92.02% on unbalance dataset and also HGB showed better accuracy of 96.31 on balance dataset with default hyperparameter tuning.

**TABLE 1.4** Performance Matrices of ML Model Results on Test Unbalanced Data

| Model | Accuracy | Precision | Recall | F1 Score | Cohen's Kappa Score | ROC–AUC Score | SD |
|---|---|---|---|---|---|---|---|
| XGB | 92.02 | 94.44 | 48.47 | 64.15 | 60.13 | 74.03 | 1.60 |
| GB | 91.70 | 93.75 | 42.85 | 58.82 | 54.60 | 71.18 | 2.76 |
| LGBM | 91.38 | 93.75 | 42.85 | 58.82 | 54.60 | 71.18 | 1.93 |
| HGB | 91.27 | 93.33 | 40 | 56 | 51.70 | 69.75 | 2.01 |
| AB | 89.04 | 76 | 54.28 | 63.33 | 58.16 | 75.65 | 2.52 |
| LR | 88.93 | 57.14 | 34.28 | 42.85 | 35.70 | 64.90 | 2.74 |
| RF | 88.93 | 75 | 85.71 | 15.38 | 12.73 | 54.03 | 1.60 |
| SVM | 88.40 | 0 | 0 | 0 | −0.83 | 49.75 | 1.30 |
| KNN | 87.66 | 50 | 57.14 | 10.25 | 7.44 | 52.36 | 1.36 |
| BNB | 85.21 | 43.18 | 54.28 | 48.1 | 37.83 | 70.92 | 3.85 |
| DT | 85.00 | 42.85 | 34.28 | 38.1 | 28.70 | 63.16 | 4.05 |
| GNB | 84.57 | 40.54 | 42.85 | 41.66 | 31.17 | 65.95 | 2.48 |

**TABLE 1.5** Performance Matrices of ML Model Results on Test Balanced Data

| Model | Accuracy | Precision | Recall | F1 Score | Cohen's Kappa Score | ROC–AUC Score | SD |
|---|---|---|---|---|---|---|---|
| HGB | 96.31 | 96.48 | 96.48 | 96.48 | 93.11 | 96.56 | 0.88 |
| LGBM | 96.62 | 96.48 | 96.48 | 96.48 | 93.11 | 96.55 | 0.73 |
| RF | 96.27 | 95.94 | 94.97 | 95.45 | 91.14 | 95.56 | 1.35 |
| GB | 94.65 | 95.50 | 95.98 | 95.74 | 91.64 | 95.82 | 1.03 |
| XGB | 95.14 | 95 | 95.47 | 95.24 | 90.66 | 95.33 | 0.75 |
| SVM | 93.24 | 94.5 | 94.97 | 94.74 | 89.68 | 94.84 | 1.68 |
| AB | 90.16 | 91.13 | 92.96 | 92.04 | 84.27 | 92.16 | 2.06 |
| LR | 86.54 | 86.2 | 90.95 | 88.51 | 76.92 | 88.51 | 2.48 |
| GNB | 85.62 | 82 | 86.93 | 84.4 | 68.57 | 84.33 | 2.08 |
| KNN | 86.17 | 79.2 | 99.5 | 88.2 | 74.1 | 87.25 | 1.6 |
| BNB | 85.86 | 85.3 | 87.44 | 86.35 | 72.97 | 86.51 | 2.54 |
| DT | 84.76 | 86.63 | 87.94 | 87.3 | 74.93 | 87.48 | 3.51 |

The best-performing ML model on both the unbalanced and balanced dataset and their corresponding confusion matric and ROC–AOC curve is presented in Figures 1.4 and 1.5.

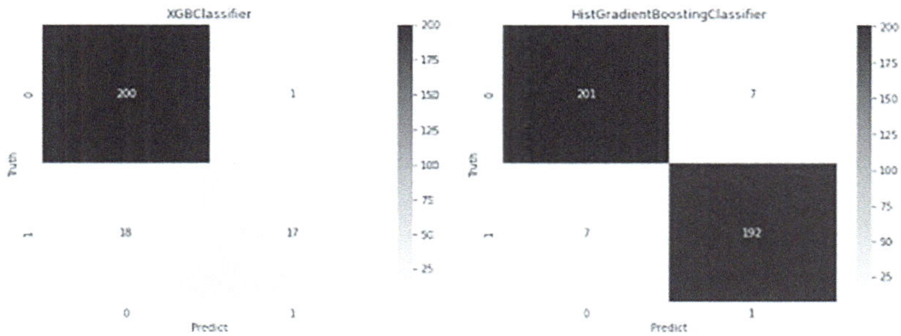


**FIGURE 1.4** Confusion matrix of XGB and HGB model on both datasets.

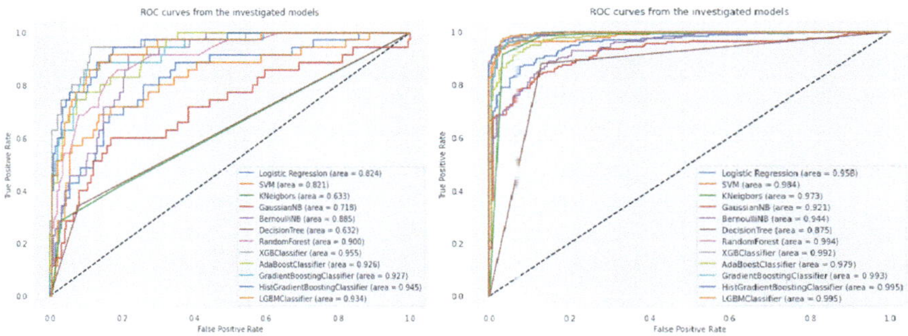


**FIGURE 1.5** ROC–AUC curve of XGB and HGB model on both datasets.

### 1.4.4 *TUNING MODEL*

While developing any model, there are chances that the model might not perform up to its full potential and thereby resulting in a suboptimal solution. Tuning helps us in unlocking the maximum potential of any model for a given problem without overfitting or deviating much from the actual results. A model is tuned by changing and selecting the best hyperparameter values that give the optimal results for a given problem. Selecting correct hyperparameters is very important as it highly influences the outcome of the model. The hyperparameter values are not something that we can predict. However, we may acquire these values from previous similar problems or using hit and trial methods, but by doing this there is a high chance of getting suboptimal results. Therefore, hyperparameter optimization techniques like Random Search CV or Grid Search CV are applied on the models for which the best-suiting hyperparameters are to be found. In this technique, the best fitting hyperparameter values are chosen from a pool of values which was provided to a hyperparameter optimization technique.

So, we thought of tuning our ML models using Random Search CV to get the best performance out of our models. As discussed previously, we too provided a pool of values to this technique from which it randomly selects different combinations of parameters and trains the model using those sets of parameters and record its results. In the end, it returns the set of hyperparameters which yields the best result. After tuning all the models, we obtained the hyperparameter values as in Table 1.6.

**TABLE 1.6** Hyperparameters After Tuning

| Models | Hyperparameters |
|---|---|
| Logistic regression | {'solver': 'liblinear', 'penalty': 'l1', 'C': 0.1} |
| *K*-nearest neighbors | {'weights': 'uniform', 'n_neighbors': 8, 'metric': 'manhattan'} |
| Random forest | {'n_estimators': 150, 'min_samples_split': 3, 'min_samples_leaf': 1, 'max_features': 'log2', 'max_depth': 12, 'criterion': 'gini'} |
| Decision tree classifier | {'min_samples_leaf': 10, 'max_depth': 10, 'criterion': 'gini'} |
| Support vector machine | {'kernel': 'rbf', 'gamma': 0.1, 'C': 10} |
| Extreme gradient boosting classifier | {'min_child_weight': 1, 'max_depth': 15, 'learning_rate': 0.1, 'gamma': 0.0, 'colsample_bytree': 0.3} |
| Adaptive boosting classifier | {'n_estimators': 200, 'learning_rate': 0.1} |
| Gradient boosting classifier | {'subsample': 0.5, 'n_estimators': 1000, 'max_depth': 7, 'learning_rate': 0.1} |
| Histogram gradient boosting classifier | {scoring = 'f1_micro', random_state = 0, max_iter = 400, max_depth = 50, learning_rate = 0.3, l2_regularization = 0} |
| Light gradient boosting classifier | {'lambda_l1': 4.642350800200348e-07, 'lambda_l2': 0.0016049159794705925, 'num_leaves': 219, 'feature_fraction': 0.5067625451535447, 'bagging_fraction': 0.5737271342724937, 'bagging_freq': 5, 'min_child_samples': 5} |

The results of the tuned model using best parameter of all the above selected ML model are represented in Table 1.7. The corresponding confusion matric and ROC–AUC curve of the best-performing model of GB with a highest accuracy of 97.48% and AUC score of 0.97 is presented in Figure 1.6.

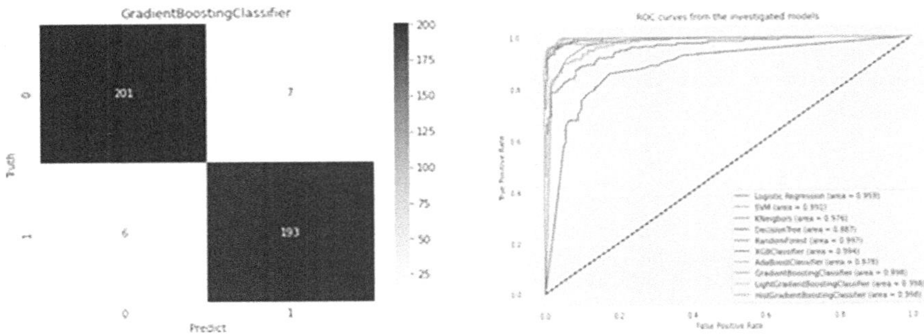


**FIGURE 1.6** Confusion matrix and ROC–AUC curve of a tuned model on test dataset.

**TABLE 1.7** Performance Matrices of Tuned ML Model Results on Test Balanced Data

| **Tuned Model** | **Accuracy** | **Precision** | **Recall** | **F1 Score** | **Cohen Kappa** | **ROC–AUC Score** | **STD** |
|---|---|---|---|---|---|---|---|
| GB | 97.48 | 96.5 | 96.98 | 96.74 | 93.61 | 96.81 | 0.89 |
| LGBM | 96.99 | 97.47 | 96.98 | 97.23 | 94.59 | 97.3 | 0.64 |
| HGB | 96.99 | 96.46 | 95.98 | 96.22 | 92.62 | 96.31 | 0.64 |
| XGB | 96.12 | 96.04 | 97.49 | 96.76 | 93.61 | 96.82 | 0.72 |
| RF | 95.51 | 96.97 | 96.48 | 96.73 | 93.61 | 96.8 | 0.73 |
| SVM | 94.6 | 99.46 | 92.46 | 95.83 | 91.12 | 96 | 1.4 |
| AB | 89.36 | 89.32 | 92.46 | 90.86 | 81.82 | 90.94 | 1.41 |
| KNN | 87.34 | 80.74 | 99 | 88.94 | 76.03 | 88.2 | 2.05 |
| LR | 86.91 | 86.41 | 89.45 | 87.90 | 75.93 | 88 | 2.07 |
| DT | 81.44 | 84.3 | 81 | 82.56 | 66.54 | 83.24 | 2.3 |

### *1.4.5 STACKING*

After getting the performance of all the tuned models on the balanced data as shown in Table 1.8, we made two stacking ensemble model one with GB as metalearner and another with LGBM as our metalearner as they were two of our best-performing models. We have taken RF, XGB, SVM, KNN, DT, AdaBoost, LGBM, and HGB as the first set of base learners for GB stacking model and RF, XGB, SVM, KNN,

DT, AdaBoost, GB, and HGB as a second set of base learners for our LGBM stacking model and then evaluated their performances as shown in Table 1.8. From both the results, we have seen that the stacking ensemble model with LGBM was performing better than the gradient boosting model, so we chose it to be our proposed model. The corresponding confusion matrices and ROC–AUC curve is plotted in Figures 1.7 and 1.8.

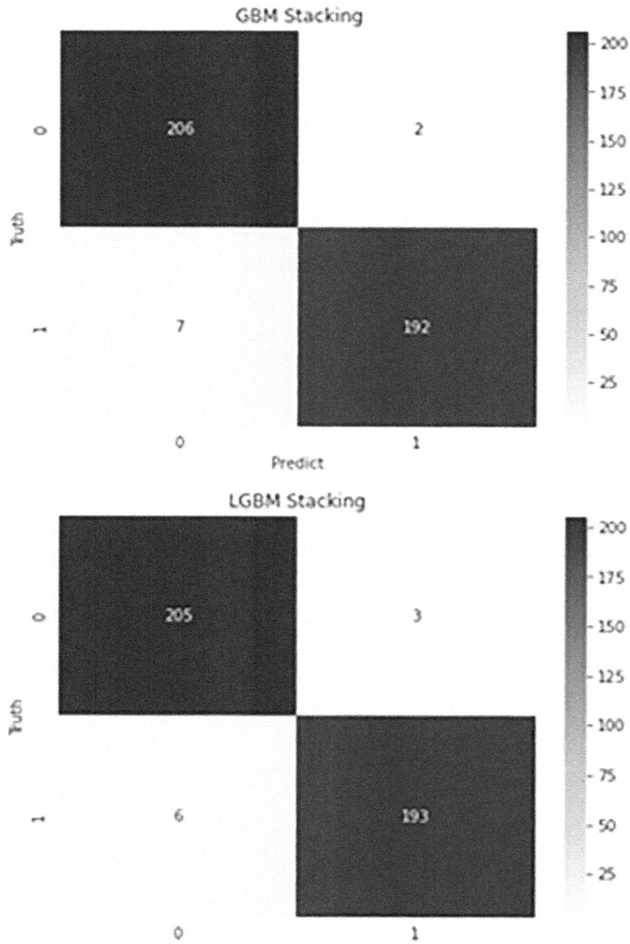


**FIGURE 1.7** Confusion matrix of GBM and LGBM stacking model.

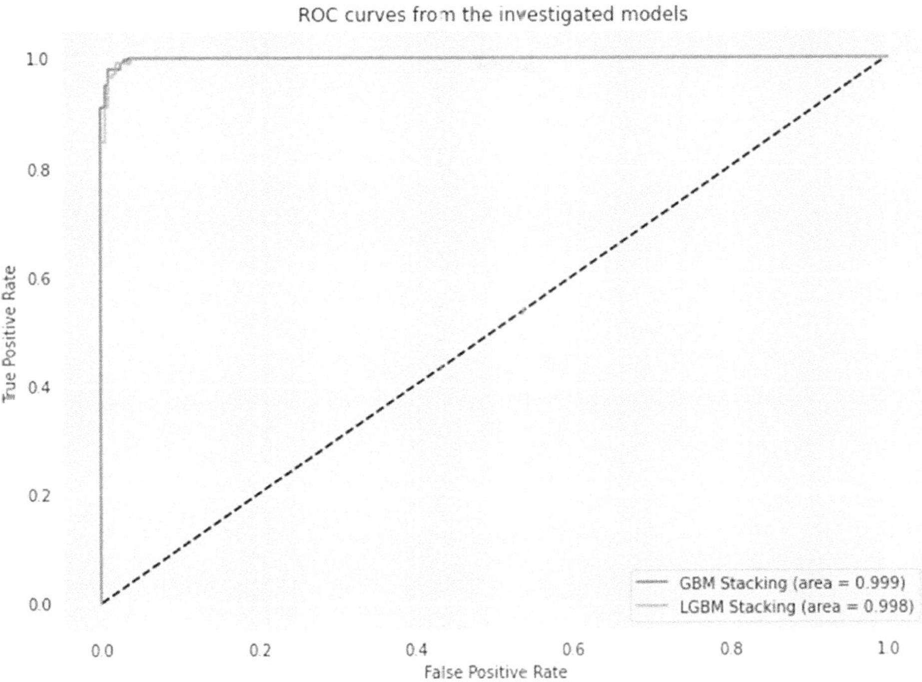


**FIGURE 1.8** ROC–AUC curve of GBM and LGBM stacking model.

**TABLE 1.8** Performance Matrices of Tuned ML Model Results on Test Balanced Data

| Meta Learner | Accuracy | Precision | Recall | F1 Score | Cohen's KappaScore | ROC–AUC Score | STD |
|---|---|---|---|---|---|---|---|
| LGBM | 98.22 | 98.47 | 96.98 | 97.72 | 95.57 | 97.77 | 0.64 |
| GB | 97.91 | 98.97 | 96.48 | 97.71 | 95.57 | 97.76 | 0.87 |

## 1.5 COMPARATIVE ANALYSIS

Table 1.9 presents the result analysis of existing literature in terms of claimed results by the authors in their proposals. Moreover, Table 1.10 presents the comparative analysis of our proposal with literary counterparts.

**TABLE 1.9** Result Analysis of Existing Literature

| References | Algorithm(s) Used | Claimed Results |
|---|---|---|
| [1] | XGB, LASSO regression | High AUC, wide net benefit threshold range |
| [23] | KNN, LR, MLP, DT, LDA | Higher precision, higher accuracy, higher $F_1$ Score, higher recall, higher AUROC |
| [24] | SVM, LDA | Higher accuracy and high precision, random forest |
| [25] | LGBM | High Accuracy and Specificity |
| [26] | LR, balanced RF | High recall rate for the minority class |

**TABLE 1.10** Comparative Analysis of Existing Literature

| Ref. | Accuracy | Precision | Recall/ Sensitivity | F1 Score | Specificity | ROC–AUC Score | Cohen's Kappa |
|---|---|---|---|---|---|---|---|
| [1] | 76.00 | | 87.00 | | 72.00 | 0.85 | |
| [23] | 94.40 | 96.40 | 91.10 | 93.70 | | 0.93 | |
| [24] | 77.00 | | 69.00 | | | | |
| [25] | 80.23 | | 74.85 | | 83.27 | 0.81 | |
| [26] | | | 37.17 | | | 0.85 | |
| Pro posed | 98.22 | 98.47 | 96.98 | 97.72 | | 0.98 | 95.57 |

## 1.6 CONCLUSION

In this chapter, we have proposed a stacking model which will be able to classify the ICU mortality of heart-failure patients with high accuracy as our model achieved outstanding performance with results such as *K*-fold mean accuracy of 98.22% on 10-fold CV and ROC–AUC score of 0.98. Our modelwas built on a dataset taken from Kaggle which was taken from MIMIC-III database. It had a total of 1177 patient instances and 51 features. Our study consisted of a total of 12 ML algorithms from which we chose nine algorithms for building our stacking ensemble classifier. Those nine algorithms were chosen after conducting multiple comprehensive sets of studies on the results of all algorithms. Our stacking classifier had LGBM as level 1 or metalearner and SVM, AB, GB, DT, KNN, HGM, RF, and XGB as level 0 or base learners. Our proposed stacking classifier outperforms the majority of benchmarking scores and ML models that were previously proposed in this study.

## KEYWORDS

- **in-hospital mortality**
- **heart-failure patients**
- **machine learning**
- **heterogeneous dataset**
- **retrospective study**

## REFERENCES

1. Li F, Xin H, Zhang J, Fu M. Prediction model of in- hospital mortality in intensive care unit patients with heart failure: Machine learning-based, retrospective analysis of the MIMIC-III database. *BMJ Open* 2021;11:e044779.
2. Amin Morid M, Liu Sheng OR, Abdelrahman S. A patient-based predictive modeling framework for early ICU mortality prediction. *ArXiv*;2017;10.
3. Pollard TJ, Mark RG, Johnson AEW. Reproducibility in critical care: A mortality prediction case study. *PMLR*;2017;68:361–376.
4. Arzenoa NM, Lawson KA, Duzinski SV, Vikaloa H. Designing optimal mortality risk prediction scores that preserve clinical knowledge. *Journal of Biomedical Information.* 2015;56:145–156.
5. Liu J, Chen XX, Fang L, Li JX. Mortality prediction based on imbalanced high-dimensional ICU big data. *Computers in Industry.* 2018;98:218–225.
6. Andrew Taylor R, Pare JR, Venkatesh AK Mowafi H. Prediction of in-hospital mortality in emergency department patients with sepsis: A local big data-driven, machine learning approach, *Academic Emergency Medicine*, 2016;23:269–278.
7. Nielsen AB, Nielsen AP, Kaas-Hansen BS, Schierbeck J. Dynamic and explainable machine learning prediction of mortality in patients in the intensive care unit: A retrospective study of high-frequency data in electronic patient records. *The Lancet Digital Health*, 2020;2:e179–e191.
8. Alves T, Laender A, Veloso A, Ziviani N. Dynamic prediction of ICU mortality risk using domain adaptation. *2018 IEEE International Conference on Big Data (Big Data).* 2018;1328–1336.
9. Anand RS, Stey P, Jain S, Biron DR. Predicting mortality in diabetic ICU patients using machine learning and severity indices. *AMIA Summits on Translational Science Proceedings*, 2018;2017:310–319.
10. Alpaydin E. Combining pattern classifiers: Methods and algorithms. In Kuncheva LI, AlpaydinE (Eds.), *Combining Pattern Classifiers: Methods and Algorithms*. 2nd edn. Wiley: Hoboken, 2014, Vol. 18, No. 3.
11. Slubana B, Lavrač N. Relating ensemble diversity and performance: A study in class noise detection. *Neurocomputing*, 2015;160(C):120–131.
12. Cortes C, Vapnik V. *Support-vector Networks*, 1995.

13. Breiman L. *Random forests, Machine Learning*. 2001;45:5–32.
14. Quinlan JR. Induction of decision trees. 1986;1:81–106.
15. Cramer JS. *The Origins of Logistic Regression*, Tinbergen Institute Working Paper No. 2002-119/4. 2002.
16. Hand DJ, Yu K. Idiot's bayes—not so stupid after all? *International Statistical Review*, 2001;69:385–398.
17. Anava O, Levy K. K*-nearest neighbors: From global to local. *Neural Information Processing Systems*. 2016;4923–4931.
18. Freund Y, Schapire RE. A short introduction to boosting. *Journal of Japanese Society for Artificial Intelligence*. 1999;14(5):771–780.
19. Friedman JH. Greedy function approximation: A gradient boosting machine. *The Annals of Statistics*, 2001;29:1189–1232.
20. Ke G, Meng Q, Finley T, Wang T. LightGBM: A highly efficient gradient boosting decision tree. *31st Conference on Neural Information Processing Systems (NIPS 2017),* Long Beach, CA, USA 2017, p. 30.
21. Guryanov A. Histogram-based algorithm for building gradient boosting ensembles of piecewise linear decision trees, Analysis of Images, Social Networks and Texts. AIST 2019. 2019;11832.
22. Chen T, Guestrin CE. XGBoost: A scalable tree boosting system. *KDD '16: Proceedings of the 22nd ACM SIGKDD International Conference on Knowledge Discovery and Data Mining*. 2016, pp. 785–794.
23. El-Rashidy N, El-Sappagh S, Abuhmed T, Abdelrazek S, El-Bakry HM. Intensive care unit mortality prediction: An improved patient-specific stacking ensemble model. *IEEE Access*, 2020;8:133541–133564.
24. Ahmad F, Ayub H, Liaqat R, Ali Khan A. Mortality prediction in ICU patients using machine learning models. 2021 International Bhurban Conference on Applied Sciences and Technologies (IBCAST) 2021, pp. 372–376.
25. Chen T, Xu J, Ying H, Chen X. Prediction of extubation failure for intensive care unit patients using light gradient boosting machine. *IEEE Access*, 2019;7:150960–150968.
26. Li L, Liu G. In-hospital mortality prediction for ICU patients on large healthcare MIMIC datasets using class imbalance learning. 2020 5th IEEE International Conference on Big Data Analytics (ICBDA). 2020, pp. 90–93.
27. Shahane S. In hospital mortality prediction, machine learning-based, retrospective analysis of the MIMIC-III database. 2021. [Online]. https://www.kaggle.com/saurabhshahane/in-hospital-mortality-prediction?select=data01.csv
28. Pedregosa F et al., Scikit-learn: Machine learning in python. *Journal of Machine Learning Research*. 2011;12:2825–2830.

## CHAPTER 2

# Predicting the Pandemic Outbreak (COVID-19) Using Facebook Prophet

KESHAV KAUSHIK

*Center for Cyber Security and Cryptology, Sharda School of Computer Science & Engineering, Sharda University, Greater Noida, India*

*Corresponding author

## ABSTRACT

The COVID-19 pandemic produced by severe acute respiratory syndrome Coronavirus 2 (SARS-CoV-2) has swept the globe, wreaking unparalleled social and economic havoc. When compared to other coronaviruses such as SARS, it has high human-to-human transmissibility. Predicting the future trajectory of the COVID-19 epidemic is critical to developing appropriate preventative and control tactics promptly. In the fight against pandemics, the ability to detect the pace where the virus spreads is crucial. Monitoring the rate of spread at random, times may aid public governments in planning human safety and policies to deal with the pandemic's effects. Recent research has proposed transmission dynamic models to better explain viral dissemination in a given population and to provide preventative actions. The COVID-19 illness features and a cumulative number of infections may be effectively analyzed using time series models. Historically, epidemic propagation has been modeled as a rate issue akin to fluid flow or heat conduction. The infection rate must be known to simulate the spread of the disease, and the percentage removal must be known to model fatalities and cures. This chapter highlights the introduction to the pandemic and the Facebook Prophet Model.

Machine Learning Applications for Data Analysis in Healthcare Systems: Sudeshna Chakraborty, Jyotsna Singh, Praveen Kumar Shukla, Prasenjit Chatterjee (Eds.)


DOI: 10.1201/9781779643193-2

The chapter will also enlighten the readers about the proposed AI-based framework to predict the pandemic.

## 2.1 INTRODUCTION TO THE PANDEMIC (COVID-19)

The recent pneumonia outbreak [1], which began in early December 2019 in Wuhan City, Hubei Region, China, is triggered by a new coronavirus (CoV) recognized by the World Health Organization as "2019-nCoV" or "2019 novel coronavirus" or "COVID-19." CoVs [2] get their title from the protruding spike proteins that give them a crown-like structure. The biology of the virus depends on spike constituents. The spike component is the part of the virus that attaches to a human cell and permits it to infect it, allowing the virus to multiply and spread to other cells. The 2019-n-Cov's exact origin, source, and biological storage are unclear, although it is suspected to be pathogenic and that bats may be to blame owing to sequence similarities to the bat-CoV [5, 13]. According to previous [3] epidemiologic studies into severe acute respiratory syndrome (SARS)- and Middle East respiratory syndrome (MERS)-CoV, the bat is the major reservoir, the palm civet or raccoon dog is the intermediate (or susceptible) host for SARS-CoV, and the dromedary camel is the dromedary camel for MERS-CoV.

When an infected person breathes virus-containing droplets and very minute particles, COVID-19 [4] spreads. Others may inhale such particles and particles, or they may resolve on their eyes, nostrils, or mouths. In some cases, they may invade surfaces they meet. Individuals being within 6 ft of an infected person are at a high risk of contracting the illness. According to current information, COVID-19 has a gestation period of 2–14 days. This happens when an individual is healthy after 14 days of interaction with a confirmed COVID-19 carrier, they will not be harmed. This chapter [5] stresses the use of blockchain in future pandemic prevention. It also covers a number of blockchain-based applications that could help the fight against COVID-19.

COVID-19 has yet to be treated with a successful treatment. The important strategies include symptomatic and supportive treatment, such as maintaining vital signs, oxygen levels, and heart rate, and addressing repercussions such as bacteremia or organ failure. The scientists [6] created a COVID-19 outbreak prediction algorithm for the top ten most densely inhabited nations. Using nine distinct machine learning techniques, the suggested prediction models project the number of new instances that are expected to develop during the next five days. The spread of COVID-19 in the top 10 affected nations is described in this chapter [7]. After China, Europe is the most infested continent. The characteristics verified, active, recovering, and mortality cases, as well as the next days' outbreak forecast, were studied in this research. State-by-state statistics from a few nations verified current instances, and death cases were also investigated.

The ability of seven well-known machine learning methods for forecasting dengue outbreaks was evaluated in this research [8]. A total of eight performance metrics are used to assess these approaches. The LogitBoost ensemble model has the highest classification accuracy of 92%, with 90% and 94% sensitivity and specificity, respectively. The very complicated structure of the COVID-19 epidemic, as well as the variance in its behavior between countries, implies that machine learning might be a useful technique for modeling the outbreak This work [9] serves as an early benchmarking exercise to show how machine learning might be used in an additional investigation. This work also implies that by combining machine learning and SEIR models, a true innovation in outbreak prediction may be accomplished.

Using a machine learning model [10], the COVID-19 epidemic in India was anticipated based on the trend seen in China. Data collected from January 22, 2020, to April 3, 2020, the model is designed to forecast the number of cases reported, recovered instances, and mortality cases. For prediction models, the time series forecasting approach is utilized. A model-based [11] forecast of the likely epidemic maximum and scale of COVID-19 in Algeria is essential to resolve the COVID-19 tough time and a devastating situation. COVID-19 management and prevention would substantially benefit from a better knowledge of the disease's dynamics. To the best of our understanding, the outbreak's unique characteristics have restricted the use of all current models. A new stochastic model is provided in this study [12] to account for COVID-19's unique transmission and prevention and capture the consequences of preventive measures adopted in Mainland China.

## 2.2 INTRODUCTION TO FACEBOOK PROPHET AND DATA VISUALIZATION

Facebook Prophet [13] is an open-source time-series modeling methodology that incorporates both traditional and novel fundamentals. It succeeds in modeling time series with multiple seasons while avoiding some of the drawbacks of other methods. The Facebook Prophet Time Series data forecasting technique has an extra restriction that accommodates nonlinear trends including yearly, monthly, and daily seasonality, as well as holiday effects. It usually performs with time series that have a lot of seasonal variation and historical information from multiple seasons. The Facebook Prophet is used in a wide range of Facebook apps to provide correct estimates for planning, control, and goal setting. It is accommodating of incomplete information and trend modifications, and it generally manages outliers. The Prophet method gives customers a lot of options for tweaking and adjusting predictions.

In this chapter, the author has evaluated a Facebook Prophet Model for the prediction of a pandemic outbreak. The author has taken a COVID-19 dataset and visualized it first to find out the country-wise plot for the number of COVID-19 cases in 2020 (Figure 2.1). Further visualization is done to find out the number of tests conducted and the number of deaths (Figure 2.2).

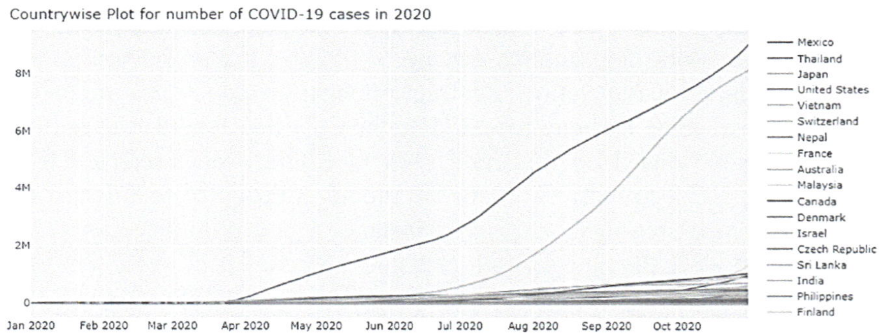


**FIGURE 2.1** Country-wise plot for the number of COVID-19 cases.

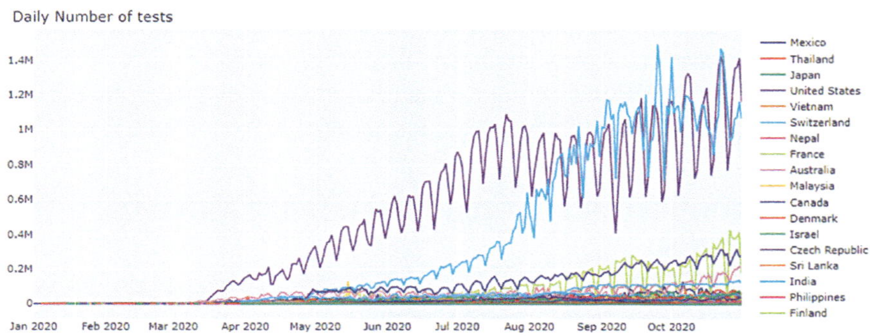


**FIGURE 2.2** Daily number of COVID-19 tests.

Figure 2.3 shows the plot for the number of deaths due to COVID-19, it has been observed that the number of deaths in the United States was skyrocketing.

Such types of visualization helped government agencies to frame the policies and take some precautionary measures against the pandemic.

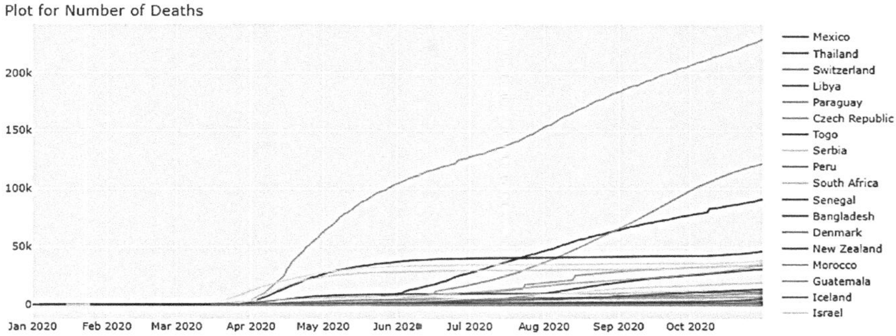


**FIGURE 2.3** Number of deaths due to COVID-19.

## 2.3 BUILDING THE AI BASED FRAMEWORK TO PREDICT PANDEMIC

After data visualization, the Facebook Prophet Model is trained. The author has forecasted for 31 days, so the data frame contains dates, which includes the training. In Figure 2.4, the plot of predicted data is plot for India for the duration of March 2020 to November 2020. As you can see in this plot, the number of COVID-19 cases increased drastically from March 2020 to November 2020.

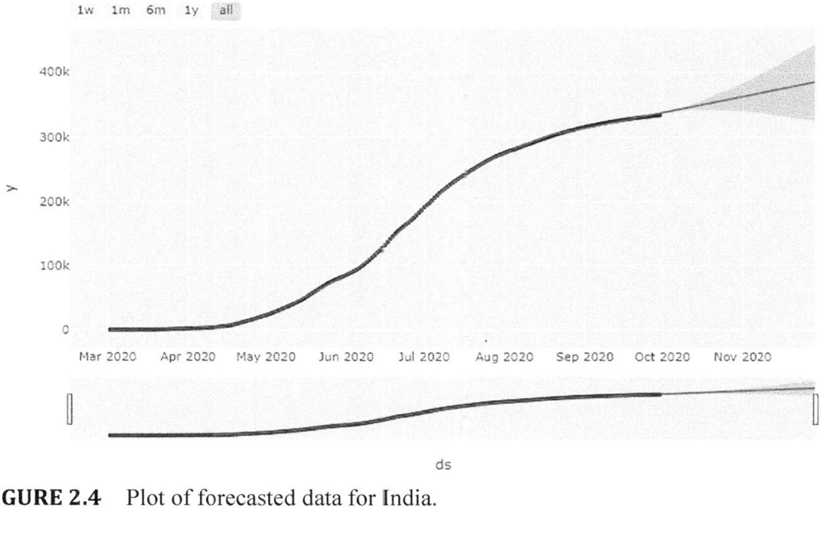


**FIGURE 2.4** Plot of forecasted data for India.

Thereafter, the change points are plotted for visualizing the predicted data in Figure 2.5. This feature assists us in identifying trend modifications that the model has inferred.

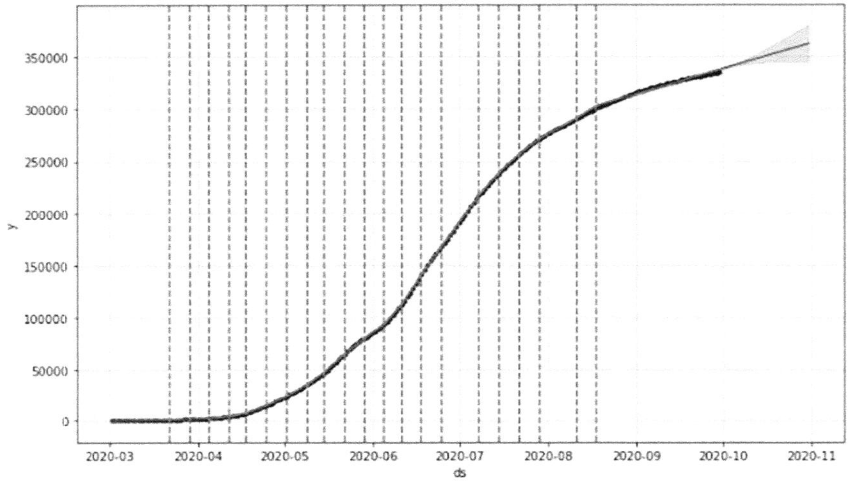


**FIGURE 2.5** Plot of change points to the plot.

## 2.4 ANALYZING THE PERFORMANCE OF TRAINED MODEL

At last, the performance of the trained model is analyzed and the original versus predicted values are plotted. The trained model is able to generate predictions for the next 31 days and for the next 60 days as well. In Figure 2.6, the comparative analysis of the model predictions versus the ground truth, that is, actually what actually happened in real life is done.

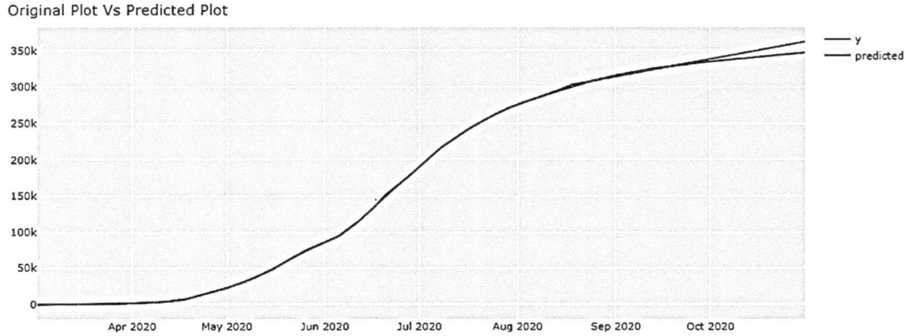


**FIGURE 2.6** Original plot versus predicted plot.

The only problem with the data that is used in this chapter is that we do not have a lot of historical information like the CoV pandemic started around March 2020. So all the information that we have from March till October or November is quite limited.

## 2.5 CONCLUSION

The spread of the CoV has brought civilization to the verge of extinction. Moreover, future transmission growth and distribution events must be researched and forecasted. Machine learning methods are crucial in the investigation and forecasting of pandemics. Machine learning techniques may also help in the detection of epidemic patterns. In this chapter, the author has discussed the rise of outbreak (COVID-19) and how the Facebook Prophet Model helped in the prediction of the outbreak. The chapter will be helpful for machine learning enthusiasts, healthcare professionals, PhD scholars, researchers, and students.

## KEYWORDS

- **coronavirus**
- **COVID-19 epidemic**
- **Facebook Prophet**
- **severe acute respiratory syndrome**
- **Middle East respiratory syndrome**

## REFERENCES

1. Fong SJ, Dey N, Chaki J. An introduction to COVID-19. In *Artificial Intelligence Coronavirus Outbreak*. 2021; p. 1, https://doi.org/10.1007/978-981-15-5936-5_1.
2. Basics of COVID-19. CDC. https://www.cdc.gov/coronavirus/2019-ncov/your-health/about-covid-19/basics-covid-19.html (accessed Apr. 12, 2022).
`3. Wu YC, Chen CS, Chan YJ. The outbreak of COVID-19: An overview. *Journal of the Chinese Medical Association* 2020;83(3):217–220, https://doi.org/10.1097/JCMA.0000000000000270.

4. Bhatraju PK et al. Covid-19 in critically Ill patients in the Seattle region—case series. *The New England Journal of Medicine* 2020;382(21):2012–2022, https://doi.org/10.1056/NEJMOA2004500.
5. Kaushik K, Dahiya S, Singh R, Dwivedi AD. Role of blockchain in forestalling pandemics. In: *Proceedings—2020 IEEE 17th International Conference on Mobile Ad Hoc and Smart Systems, MASS 2020*, December 2020, pp. 32–37, https://doi.org/10.1109/MASS50613.2020.00014.
6. Khakharia A et al. Outbreak prediction of COVID-19 for dense and populated countries using machine learning. *Annals of Data Science* 2021;8(1):1–19, https://doi.org/10.1007/S40745-020-00314-9/TABLES/3.
7. Yadav D, Maheshwari H, Chandra U. Outbreak prediction of covid-19 in most susceptible countries. *Global Journal of Environmental Science and Management* 2020;6(Special Issue (Covid-19)):11–20, https://doi.org/10.22034/GJESM.2019.06.SI.02.
8. Iqbal N, Islam M. Machine learning for dengue outbreak prediction: A performance evaluation of different prominent classifiers. *Informatica* 2019;43(3):363, https://doi.org/10.31449/INF.V43I3.1548.
9. Ardabili SF et al. COVID-19 outbreak prediction with machine learning. *Algorithms 2020* 2020;13(10):249, https://doi.org/10.3390/A13100249.
10. Tiwari S, Kumar S, Guleria K. Outbreak trends of coronavirus disease—2019 in India: A prediction. *Disaster Medicine and Public Health Preparedness* 2020;14(5):e33–e38, https://doi.org/10.1017/DMP.2020.115.
11. Bentout S, Chekroun A, Kuniya T. Parameter estimation and prediction for coronavirus disease outbreak 2019 (COVID-19) in Algeria. *AIMS Public Health* 2020;7(2):306, https://doi.org/10.3934/PUBLICHEALTH.2020026.
12. Zhang Y, You C, Cai Z, Sun J, Hu W, Zhou X-H. Prediction of the COVID-19 outbreak based on a realistic stochastic model. *medRxiv* 2020; , https://doi.org/10.1101/2020.03.10.20033803.
13. Time Series Analysis with Facebook Prophet: How it works and How to use it | by Mitchell Krieger. Towards Data Science. https://towardsdatascience.com/time-series-analysis-with-facebook-prophet-how-it-works-and-how-to-use-it-f15ecf2c0e3a (accessed Apr. 12, 2022).

CHAPTER 3

# Mobile App for Analyzing and Predicting Depression in Human Beings

VIRENDRA KUSHWAH, PRACHI BHATT, TOSHINI AGRAWAL, JIGYASA BISHT, and SHIVANGI SINGH

*School of Computing Science Engineering and Artificial Intelligence, VIT Bhopal University, Bhopal, Madhya Pradesh, India*

*Corresponding author

## ABSTRACT

Depression is a widespread mental health issue affecting over 264 million people worldwide. It is characterized by fluctuating mood changes and emotional responses that can be short-term or long-lasting. When untreated, depression can significantly disrupt daily functioning and reduce a person's potential, sometimes becoming as debilitating as a physical illness. Timely diagnosis and treatment are crucial for managing depression. While there are several mental health apps available, many of them have premium features, making them inaccessible to users who cannot afford the costs. One such app, Mood Path, tracks mood patterns and offers health assessments to help users manage negative thoughts, focusing on mindfulness skills rather than physical exercises. In contrast, our app takes a more comprehensive approach by analyzing user inputs and recommending tailored treatments. It is free to use, serving as an accessible eHealth tool for individuals dealing with depression. The app emphasizes treatment and recovery, offering a user-friendly interface that allows easy interaction in everyday language. To ensure the app remains beneficial, it includes scheduled treatments to help users manage their anxiety and negative thoughts. This approach minimizes the risk of overuse while promoting long-term mental health improvement.

Machine Learning Applications for Data Analysis in Healthcare Systems: Sudeshna Chakraborty, Jyotsna Singh, Praveen Kumar Shukla, Prasenjit Chatterjee (Eds.)


DOI: 10.1201/9781779643193-3

## 3.1 INTRODUCTION

Just like a therapist, this app allows the users to feel comfortable with his/her emotions by interacting with him/her and based on analysis (by asking questions—Q&A sessions) app tracks the symptoms and segregate (depends on input answers) them based on level of depression they are in finally app recommends the treatment using different methods like meditation, yoga, listening to audiobooks providing healthy diet chart, all customized to users for help in healing him/her.

### *3.1.1 MOTIVATION*

Depression has affected more than 264 million people as it has been a common illness worldwide.

Globally, people of all ages can suffer as a prominent cause. Due to the numbers still increasing day-by-day, it has become a contributor to the overall disease index worldwide. As, there have been many causes like abuse, age factor, death in the family, certain medications which may lead a person to attempt suicide. Around 8 lakh people die due to suicide every year.

Suicide is the number two cause for the high rate of deaths in the age gap of 15–29-year olds. The developing countries and the underdeveloped countries face a high rate and unfortunately could not receive much help regarding that due to less knowledge and acknowledgment about it. With the help of this app, we want to help as many people as we can.

### *3.1.2 METHODOLOGY*

This app opens with the logo of our app, that is, "worry not" and later shows a motivational quote to motivate the users. After the motivation quote comes to the login page where the user inputs "name," "email address," and "password." This is done so that the app can differentiate between users and this way the user will have their data secured. After the login page comes the questionnaire pages where the app asks the user certain questions that will help in the detection of the severity of the user's depression. The determination of the level of depression is done through a point (here "stars") system, that is, each answer the user gives will have certain points allocated to them.

In our app, we have made a point (or star) system such that if the user gets a point (or star) ranging from 0 to 3 then the user will be detected to have mild depression, if the user gets points (or star) ranging from 4 to 9 then the user has intermediate depression and if a user gets points (or stars) more than 9 then the user has extreme depression. This point (or star) system is not visible to the user so as not to affect their mood. After the determination of the level of depression by our app, the app recommends the user tasks like listen to songs, or do meditation, and so on. This helps the user to take their mind off the depressive or sad thoughts they have in the present time. Also, as mentioned in paper [5], apps should be friendly and colorful so they can enjoy using them and also more affordable than the books that help you improve depression and our app once done will be tried to be made free of cost. Our app also recommends users the personalized diet chart that can keep them healthy if the users follow it.

### 3.1.3 *OBJECTIVES OF THE WORK*

- Determine the level of depression of the user.
- Recommend them tasks that will help them get through depression.
- Proper check on the eHealth of the users.
- Reduce their stress.

## 3.2 RELATED WORKS

"The Digital Psychiatrist": where they proposed the accurate results that should come out of anxiety and depression-based apps and clearly dictated that they should be able to focus on one disorder rather than a collective which is the case in our app as it purely focuses on the depression and curing it. The limitations of this study include the use of terms for the experiment which may not particularly reflect what the user might use. For example, the user might use "anxiety," "depression," or "worry" rather than "interpersonal therapy." The second limitation is the study has some anomalies as the description might not necessarily be true with the apps functioning.

"Finding a Depression App: A Review and Content Analysis of the Depression App Marketplace" in which they mentioned the coding if the main purpose should be categorized into self-assessment, symptom monitoring, psychoeducation, and so on. They mentioned that some apps which provide therapeutic treatment audio use multimedia for the purpose of hypnosis treatment in depression but the audio treatment provided by our app is for the sole purpose of listening and then relaxing, keeping the mind at peace, as it consists of audiobooks, podcasts, and soulful music for different kinds of meditative practices. Out of the 243 apps tested and examined by the writers, only 4.1% account to provide spiritual healing and entertainment that is only 10 apps so to contribute to the group our app also relatively provides fun yoga exercises and exciting audio stories to listen to. Furthermore, the limitations consist of the fact that the findings may not have represented all the apps throughout the globe because some apps just provide local areas or a restricted area. The second limitation to this app is this research only focuses on depression-based apps which leaves out the nondepression apps, as the International College of Business and Technology is one of the most common cures for depression but is mostly found in nondepression apps. In paper [7], findings show that usage of social media is directly proportional to the risk of depression means higher social media usage consequently higher will be the risk of depression and is significantly higher among teenage girls. To control the risk of depression an early detection system is proposed. In paper [8], it investigates the need for an online support forum among people with depression. The study conducts a survey among people and 134 participated to report their experience and queries like treatment and services, medication, understanding depression, and so on.

Furthermore, a paper "Health App Use Among Individuals with Symptoms of Depression and Anxiety: A Survey Study with Thematic Coding" reported that fitness and nutrition are the forte of most apps including and pinning mostly on health practices like weight management, diet chart, and different types of exercises. This paper focuses on the types of users of mental health apps rather than what apps provide to get an insight on fact of what customers expect from virtual treatment. The limitations of this research include the point that some of the participants have either taken treatment in the past or were involved during the experiment. Moreover, the demographic is of the Midwest region in the USA consisting of high-income earning females having high society background.

The paper "IntelliCare: An Eclectic, Skills-Based App Suite for the Treatment of Depression and Anxiety" talks about the "Intellicare" app which is a collection of different apps based on different types of requirements required by a person suffering from depression. It consists of "Boost Me" which provides positive activities to boost the mood of the user, similarly, iCope is where a user would find a library of recordings that consist of relaxed, mindfulness of audio to feel calm and at peace and there are many more apps like this inside of one single app. However, the limitations of it consist that results were taken from the samples which were already improving. Second, the coaching protocol was creating a conflict as it was recommending the use of all apps rather than based on frequent use by users.

In paper [6], named "Current Knowledge and Adoption of Mobile Health Apps Among Australian General Practitioners: Survey Study," they surveyed many Australian practitioners where they provided insights that most of the health apps were recommended for patients and more than 50% of them were mental health-related apps. This survey helped us motivate toward making this app. Their limitations included a very low response rate and they were election bias which is not handled clearly for accuracy.

In a paper named, "Smartphone Apps: Review and Evidence-Based Recommendations for Future Developments," they again surveyed and analyzed the approaches which can be woven into further innovative apps for mental health. The factor they mentioned is that it is appreciated if the apps are designed for use for those who have not sought treatment. So that they could get help easily and in the comfort of their home. Also, they mentioned that apps should be able to allow the user some tailoring with the schedule provided for their treatment which is why we segregated this feature.

Ekinci [9] investigates variables that are the cause of depression among Egyptian youth and a study found that the main reason for depression is anxiety and depression is significantly higher among female Egyptian youth. In comparison with American youth, the mean score on depression is higher among Egyptian youth.

In paper [10], the study is totally based on depression level prediction using a sampling method of middle school students on the basis of sports participation, gender, age, and socioeconomic status.

Additionally, there are fewer apps that are free for patients with depression-like "Happify" which has different types of games and activities which can alleviate your mood and is for positive psychology for making you feel better but just for short amounts of time. The limitations of this app include that this only focuses on mind engagement rather than physical, although it requires movement but not effectively as one would require during a particular exercise.

Furthermore, there is an app named "Relieve Depression Hypnosis" which gives visual, audio, and rhythmic tools for achieving calmness and can be used for meditation and or sleep. It helps in removing negative subconscious thoughts which limit full focus and determination. The limitations include that it only focuses on the thoughts rather than improving the physical aspect. Also, it should be known that some people might feel confident if they have a healthy lifestyle.

Moreover, another app "What's up" provides a habit tracker and not any kind of treatment which is similar to ours where ours just provide treatments on a surface level and not detailed licensed therapy. They provide pocket therapy as told by Dr. Raichback in the article 25 best apps for depression "the apps allow for privacy and confidentiality and can be safe for individuals who may be too ashamed to admit about their mental health issues." Also, as written by Promod Kerkar, M.D. on Pain Asist., the apps which work on self-healing rather than highly composed drugs. The limitations of this app include the logging problem when using the forum.

## 3.3 EXISTING SYSTEM

The existing work like the app Talk Life allows you to be a part of a support group call or be able to meet a person who would listen, talk, and help you feel less sad and lonely. The app known as Sanvello provides users with breathing exercises that can help in reducing stress, balance mood, control anxiety issues, and manage anger. Similarly, cognitive behavioral technique (CBT) apps take assessments to check your strength and weaknesses toward the taking assessment test to check strengths and weaknesses and keep a journal for it. Also, as mentioned about the journal, it can be text-based or log-based where the time slots and the meditation exercises done can be recorded to follow the improvement.

### *3.3.1 DRAWBACKS AND ISSUES IN EXISTING WORK*

A study was done on more than 100 depression apps where it was analyzed that only 35% were clinically effective and only 3% of even them were actually accurate according to clinical standards. It also points toward the negligence and lack of valid information through five-star ratings for these apps. Altogether, these results suggest that there still are serious and overly neglected challenges that can be born from these killer apps if seen from a clinical quality perspective. Improving consensus is that most commercially available apps are not evidence-based which can be highly dangerous. Each person's lookout is different for each problem when suffering through depression and they have a different front on that, a single app provides a basic front/solution on the given problem which may not be able to fulfill the purpose.

Symptom monitoring has been highly praised as helpful and supportive with people who have mental issues. Despite that fact, there have not been many apps that provide this feature and on that front also even if some do they do not give official output as accurate. As some apps could only help for not more than two weeks, an attachment and getting comfortable with these apps can be highly dangerous for the patients as they might feel weak after not getting the same environment, even if they improved while using those apps.

According to the published papers on depression apps, the factors which are causing issues with existing work are mostly the factors related to lifestyle like nutrition, stress (financially and digitally), consistent fitness routine, and diet basically routine activities that keeps a person fit from waking up in the morning to getting a good sleep at night and to improve this when they are suffering from depression is what drives them to download these apps. Lastly, apps are very expensive and consist of premium packages which are very costly for users to afford.

## 3.4 PROPOSED SYSTEM

We have designed the mobile app in a secure manner for the end-users. It provides flexibility to the user to go through various functions of apps and we had compressed a substantial number of files. Also, the CBT is a psychotherapeutic treatment where people suffering from anxiety, depression, panic attacks, or eating disorders are provided with a treatment where they identify their thoughts and patterns of negative thoughts that disturbs their mind. Some of the CBT also includes setting goals which is a key feature in our app, where while tailoring their schedule they can set their daily, weekly goals. In the beginning, we used a splash activity function. Splash Screen is one of the exciting features in Android and is referred to as a welcome screen or user's first experience of an application. Splash screens give a professional look to our system. A delay method is used in this, which fixes the on-screen time of the page. In our app, we have used a delay time of 3000 which means the first page will be active for 3 s after that it will move on to the next page/activity.

Figure 3.1 depicts the operational workflow of the proposed app model. The proposed model consists of modules such as analysis, segregation and recommendation.

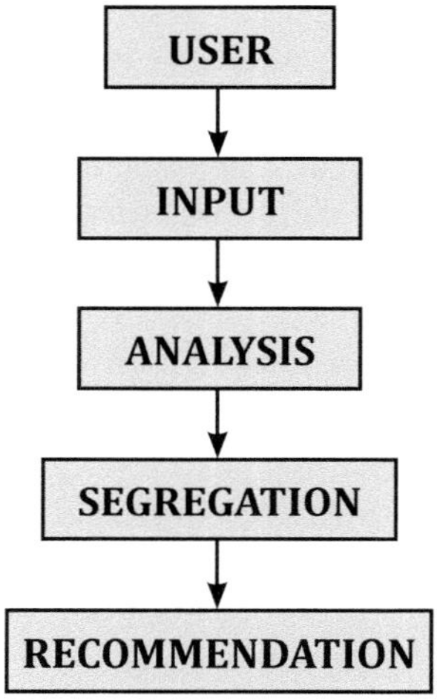


**FIGURE 3.1** Operational workflow.

Figure 3.2 shows the login page of the proposed mobile app. On the second page is our welcome page/activity, two things are used. First, the code is written in XML and we have used an image button in it. On the welcome page/activity, there is an image written click here. After clicking on that page, it will move to the next page/activity. Then in the login page/activity, we have used shared preferences. Shared Preferences gives simple methods by pointing to a file that has key-value pairs just to allow reading and writing them. The file is supported and handled by the framework and can be private or shared. We stored the ID and password of the user. Then in the questions part, we have framed the questions in the XML part and the working procedure in Java's main activity code. And here in this part, we have done the analysis so that we could know the level of depression of users based on their answers. The navigation bar is used for providing the assorted options of meditation, yoga, audiobooks, and diet charts. There these things are mentioned level wise. Videos, music in the form of audiobooks, chants, and images have been used in that page/activity. So that via images the user may feel good.

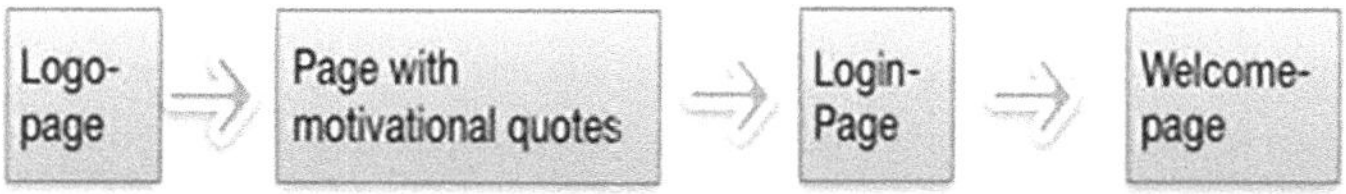


**FIGURE 3.2** Login page.

We have divided the app into three modules. Figure 3.3 depicts the query-based analysis of depression. In this stage, the questions are posted to mobile app users to identify and analyze the level of depression.

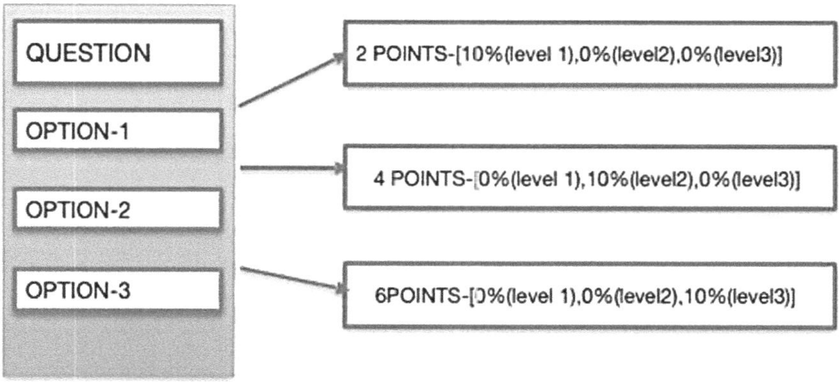


**FIGURE 3.3** Query-based analysis of depression

### 3.4.1 MODULE 1: INPUT

This module comprises four pages which include a logo page, a page that has motivational quotes, a login page, and finally a welcome page. The login page takes the data from the user. The Shared Preferences method() which was login() which can store a small amount of data. The login happens only one time for the user to feed his/her details. If the user successfully logs in then the login screen will not appear again. If the submit button is not pressed then the user will get that screen again whenever he/she will open the app.

The design of the login page has three sections which are name-mail ID and password. These need input by the keyboard.

### 3.4.2 MODULE 2: ANALYSIS

Two Java activities have been made for this module which consists of eleven questions that are to be answered by the user which is related to judging the intensity of the symptoms. In the first Java activity, the questions and answers are mentioned using get Answers(), where if the answer is chosen then it would be true and if not then false. Again, Shared Preferences() has been used, so that the same questions do not appear repeatedly. Also, a variable named answerques() was made, where if the variable is true, that means if the answer has been submitted before the time, that is, 40 s then that question will not appear again, if not, that is if the user took more than 40 s then that question screen will appear again: (1) frontend and (2) backend.

### 3.4.3 MODULE 3: SEGREGATION AND RECOMMENDATION

This module comprises the backend part where the analysis part is concluded into three stages which are mild, intermediate, and extreme, that is, level 1, level 2, and level 3. When the user answers the question, each option comprises some points which at the end are summed up and divided into three levels. If the points are between a given range, then the user would be allocated that level but he/she will not know about it. With the use of a database, it can be achieved and easily the information can be stored in the backend.

This module includes the four types of treatments divided into three levels. Fragments were made for each level of each treatment as through the layout the activity was made which was the design of the page and through Java file, the contents were executed. UI was used where Java files are made. There were two Java files for one fragment in a developer made folder. Those two files are "java_fragment" and "java_viewmodel." The latter file helps if something needs to be printed on the screen. This file is used to execute which design should be seen on the screen from the activity file. So, the execution of audios and videos is done through "java_viewmodel." Also, to make this less complex, another folder of Java has been created for, say, one of the treatments, where for each level the above-mentioned files which are "java_fragment" and "java_viewmodel" are created. Like "java_fragment_level1" and "java_viewmodel_level1."

### 3.4.3.1 AUDIO BOOKS, PODCASTS, AND SOULFUL MUSIC

In this, there are three levels created which include these things to calm the user and make him feel comfortable. Fragment_level1_audio file was created in the folder named audio, where the adjust volume option is provided through the VolumeProvider() method.

### 3.4.2.2 DIET CHART

The diet chart is simply provided through the OnclickListener() by simply clicking on the levels in the navigation bar, in which at all levels the eatables that fulfill the nutritional requirements for dinner, lunch, and breakfast are provided.

### 3.4.2.3 MEDITATION

Here the audio of diverse types of meditative practices is provided which can be a start and pause. The media player object was imported from the Android studio and then the start, stop, and pause functions were called which were defined in the media player class.

### 3.4.2.4 YOGA

Here on Create() lifecycle callback method is called where the mediaSessionCompat() and mediaControllerCompat() classes are created and stop() and pause() functions are called to play the video which includes different yoga poses and how to do them. The design includes the gradient background with the video and background music playing which are supposed to motivate the user.

## 3.5 RESULT DISCUSSION AND ANALYSIS

### *3.5.1 GENERAL DEPRESSION ANALYSIS GRAPH*

This study includes a number of cross-sectional experiments with a total of 41,344 participants. Outpatients from otolaryngology clinics had the highest prevalence of depressive symptoms (53.0%), followed by dermatology clinics (39.0%) and neurology clinics (39.0% and 35.0%). According to subgroup studies, the prevalence of depression and depressive symptoms ranged from 17.0% to 53.0% in different specialities. Outpatients from developing countries have a greater prevalence of depression and depressive symptoms than outpatients from developed countries. Furthermore, between 1996 and 2010, the prevalence of depression and depressive symptoms in outpatients decreased significantly [2]. Figure 3.4 shows a survey on depression analysis.

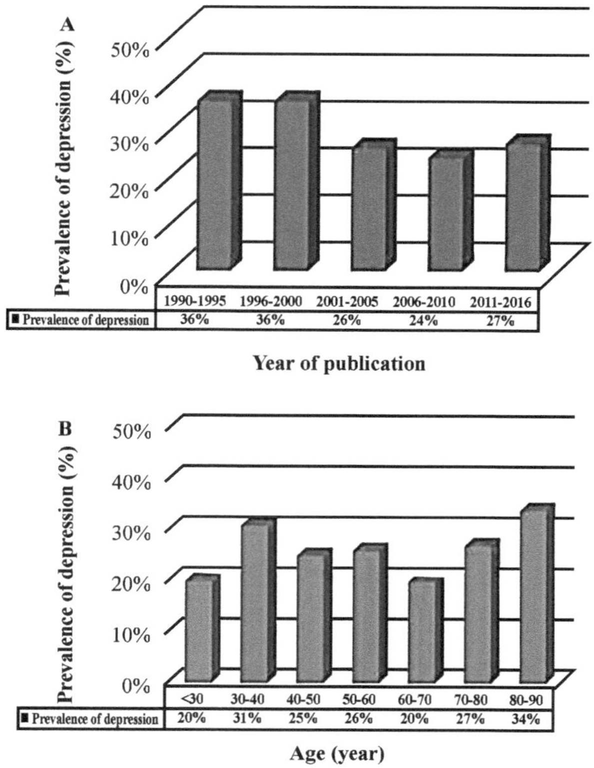


**FIGURE 3.4** Survey on depression analysis.

Figure 3.5 shows depression analysis surveys in India. This survey was done by the National Mental Health Survey for the year 2015–2016; from this study, it was shown that the age groups that suffer more from mental problems, such as depression. And a study was done on the types of mental disorders and the types that people are facing. The above graph shows us more about people's stress levels. How many of them feel depression is taboo? A compact study was done by the government in this regard [3].

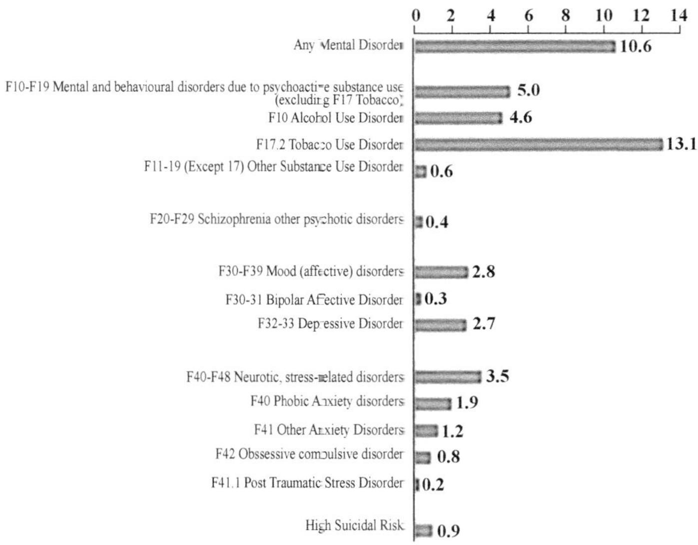


**FIGURE 3.5** Survey on depression analysis in India.

### 3.5.2 OVERALL SURVEY GRAPH OF DEPRESSION ANALYSIS IN INDIA

One of the known apps is "Moodpath: Depression and Anxiety." This app has a rating of 4.3. This app has been used by many people all over the world. Most people like this app, as it provides them with a graph of their overall performance in a week or month. It provides positive quotes and even allows other users to share their thoughts. Users dislike this app because it does not provide them with proper security. And people, moreover, provide negativity too.

Another known app is "Talk life: Depression and Anxiety." This app has a user rating of 4.3, a great success rate. According to the reviews of people, this app is really supportive in communicating with the users out there. This app mainly focuses on users' mental health and problems regarding it. Where many phase problems are in getting a proper notification from other users there and sometimes, they are not able to switch to different apps because of the data that can be lost in that app. What people mostly dislike in this app is that to use many of its functions one has to pay for it.

These were the existing works of some depression analysis apps. We went through the drawbacks of these apps and tried to make our app better by avoiding these problems. Our user just needs to log in one time, and all details will be stored. In addition, our app is well managed with all positive and motivational thoughts.

### 3.5.3 RESULT FOR EXISTING DEPRESSION ANALYSIS APP

There are many mobile apps now which are designed to overcome depression, change people's moods, and make their health life cycle. In August 2014, 2923 people were screened across five waves of two-week advertisements across the country. A total of 626 participants were randomly assigned to one of three study arms, each with an iPad 2.0 and a smartphone (iPST, 211; Project: EVO, 209; Health Tips, 206). The sample's mean age was 33.95 (SD 11.84), and the sample's mean Patient Health Questionnaire (PHQ-9) score at baseline (13.64, SD 4.95) showed mild depression. The percentage of people in our sample with a PHQ-9 total score between 5 and 10 was 23.0% (144/626), while the percentage of people with a score of 10 was 77.0% (482/626). Only 11 participants with a total PHQ-9 score of less than 5 reported a score of more than 2 on this tool's item 10, despite their baseline total PHQ-9 being 3.09 (SD 0.83) [1].

### 3.5.4 MOBILE APP—WORRY NOT

Figure 3.6 shows Worry Not user-friendly mobile app to analyze and predict depression. In paper [4], also the factors which make mental health apps work for high-efficacy apps, it says that there should be a simple user interface and they should be able to navigate through the app functioning without having to stress much. Depression in human beings is predicted using query-based analysis and levels of depression are identified in patients. Accordingly, treatments to the patients will be given.

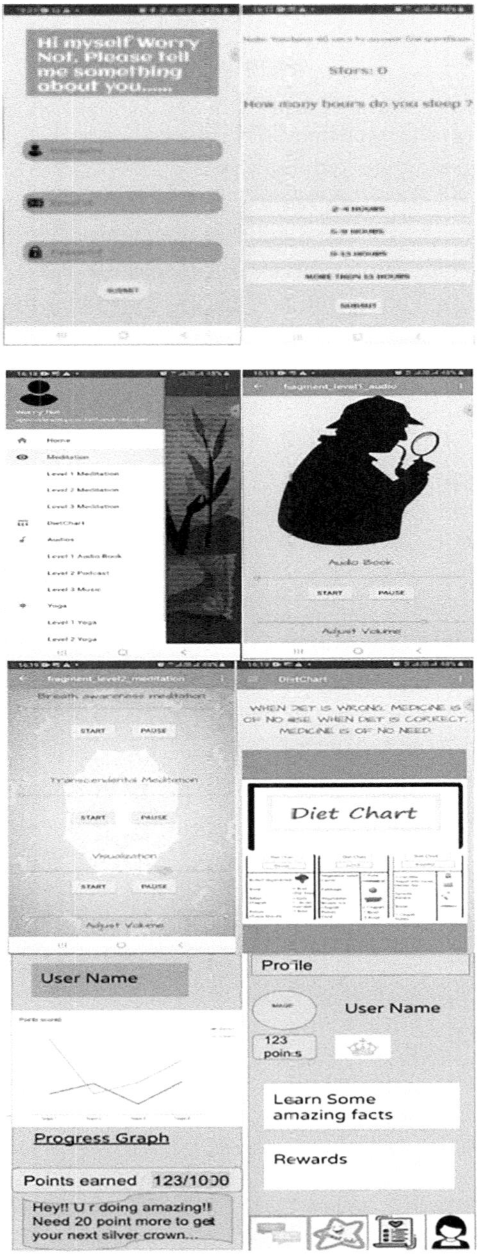


**FIGURE 3.6** User-friendly mobile app to analyze and predict depression.

## 3.6 CONCLUSION AND FUTURE WORKS

### 3.6.1 CONCLUSION

Our app is built on the fact that some people around the world could not afford to go to expensive treatments by doctors offline. So we would make it online and much more interactive so he/she does not feel that disconnected from the real process. Our app has achieved the proposed ideas that wanted to achieve the above-set objectives. We will enhance some things which will make it much more attractive and effective as well. This app has been made by keeping in mind the fact why users would use this and how beneficial it can he/she be to remove depression. These types of mobile apps are really needed in today's man-machine life scenario.

### 3.6.2 FUTURE WORK

- Producing innovative ways in which the user can play games or do activities based on which we will decide their level without them noticing.
- Graph-level monitoring of users through the treatments.
- Keep track of their change in behavior and mind perspective by providing rewards in the form of greetings.

## KEYWORDS

- **mHealth**
- **eHealth**
- **depression**
- **cognitive behavioral technique**
- **mobile app**

## REFERENCES

1. Arean PA, Hallgren KA, Jordan JT, Gazzaley A, Atkins DC, Heagerty PJ, Anguera JA. The use and effectiveness of mobile apps for depression: Results from a fully remote clinical trial. *Journal of Medical Internet Research [Electronic Resource]* 2016;18(12):e330, https://doi.org/10.2196/jmir.6482

2. Wang J, Wu X, Lai W, Long E, Zhang X, Li W, Zhu Yi, Chen C, Zhong X, Liu Z, Wang D, Lin H. Prevalence of depression and depressive symptoms among outpatients: A systematic review and meta-analysis. *BMJ Open* 2017;7(8):e017173. https://doi.org/10.1136/bmjopen-2017-017173
3. Srinivasa Murthy R. National Mental Health Survey of India 2015–2016. *Indian Journal of Psychiatry* 2017;59(1):21–26. https://doi.org/10.4103/psychiatry.IndianJPsychiatry_102_17
4. Chandrashekar P. Do mental health mobile apps work: Evidence and recommendations for designing high-efficacy mental health mobile apps. *mHealth*. 2018;4:6. https://doi.org/10.21037/mhealth.2018.03.02.
5. East ML, Havard BC. Mental health mobile apps: From infusion to diffusion in the mental health social system. *JMIR Mental Health*. 2015;2(1);e10. https://doi.org/10.2196/mental.3954.
6. Byambasuren O, Beller E, Glasziou P. Current knowledge and adoption of mobile health apps among australian general practitioners: Survey study. *JMIR Mhealth Uhealth*. 2019;7(6):e13199. https://doi.org/10.2196/13199.
7. Tong Mok W, Sing R, Jiang X, See SL. Raffles Girls' School: Singapore, Singapore, SG, Investigation of social media on depression. *The 9th International Symposium on Chinese Spoken Language Processing*. 2014. https://doi.org/10.1109/ISCSLP.2014.6936690
8. Barney LJ, Griffiths KM, Banfield MA. Explicit and implicit information needs of people with depression: A qualitative investigation of problems reported on an online depression support forum. *BMC Psychiatry*. 2011;88(2011).
9. Ekinci NE. Investigation of depression level of middle school students' according to some variables. *Asian Journal of Education and Training*. 2018;4;280–283. https://doi.org/10.20448/journal.522.2018.44.280.283
10. Ghareeb Ghareeb. *Conference: The World Federation for Mental Health At: Cairo World Congress, Cairo An Investigation of Some Variables Related to Depression in Egyptian Youth*. 1987.

# CHAPTER 4

# Machine Learning-Based Analytics for Premature Rheumatoid Arthritis and Osteoarthritis Detection in Clinical Practices—A Review

GANESH KUMAR M. and AGAM DAS GOSWAMI*

*School of Electronics Engineering, VIT-AP University, Amaravati, Andhra Pradesh, India*

*Corresponding author

## ABSTRACT

People experience significant health discomfort as a result of rheumatoid arthritis (RA) and osteoarthritis (OA). There are numerous other types of arthritis that have an impact on people's comfort and efficiency. In terms of using computer-aided diagnostic solutions, previous research has concentrated on a specific set of digital imaging solutions and predictive analysis models for the accurate identification of issues. However, with the increased use of machine learning (ML) models, the scope of more accurate predictions has expanded. Some of the key models discussed in the studies are detailed in this chapter which focuses on the literature pertaining to RA and OA. In terms of understanding the scope of research, the discussion section presents the gaps found in the literature, as well as how certain key aspects, such as the application of lifestyle factors and the use of a more accurate detection process for correctly classifying arthritis, are some of the significant areas for research. The goals of future research are defined as focusing on a possible set of evolutionary algorithms and feature extractions that can be useful for improving overall accuracy.

Machine Learning Applications for Data Analysis in Healthcare Systems: Sudeshna Chakraborty, Jyotsna Singh, Praveen Kumar Shukla, Prasenjit Chatterjee (Eds.)


DOI: 10.1201/9781779643193-4

## 4.1 INTRODUCTION

Artificial intelligence (AI) computing has altered the dynamics of development and is playing an important role in improving people's quality of life. Healthcare solutions are one such domain where the scope of application for AI has become an urgent need. There are distinct sets of solutions where AI's role is prominent, ranging from formulation planning to research to clinical care.

Predictive modeling solutions are becoming more constructive for some of the current AI applications in the healthcare domain, from diagnosis to treatment care, and they help improve the accuracy of diagnosis and treatment. Rheumatic or musculoskeletal conditions cause more than 150 diseases, and the syndrome is usually progressive and painful for those who suffer from it. Rheumatic conditions can be divided into three categories: joint diseases, spinal disorders, and conditions that lead to traumatic conditions. In general, such rheumatic conditions cause morbidity and disability in certain conditions, as well as significant healthcare expenditures [1].

In terms of the dynamics of applying machine learning (ML) models, one of the key areas of healthcare where the use of ML models can be seen as essential is the accurate prediction of rheumatoid arthritis (RA).

RA is a chronic system disease that affects joints, fibrous tissues, connective tissues, and other tissues, according to a brief overview of the RA and osteoarthritis (OA) outline. It primarily affects people between the ages of 20 and 40, during their most productive years of adulthood. More often than not, if undiagnosed for an extended period of time, it can impair the healthy movement of people with RA [2].

OA is a degenerative joint disease that primarily affects the articular cartilage. More often than not, the problem manifests itself in the cases of aged people and physically stressed people as a result of excessive physical work. According to statistics, OA is one of the 10 most disabling diseases in developed countries [3, 4].

The fact remains that there are various types of arthritis, with OA and RA being two of the most common. Because the symptoms of both types of arthritis discussed are similar, there is a need for more accurate and appropriate treatment to decide on the best treatment to address the issues [5].

Knee osteoarthritis (KOA) is a condition of the human knee joint that primarily affects the cartilage. Cartilage plays an important role in leg movement. The top layer of cartilage disintegrates and deteriorates in OA, resulting in excruciating pain. OA is a chronic joint disease that causes pain, swelling, stiffness, and loss of function. OA is a chronic disease whose origin is unknown.

It is the most common type of arthritis and is distinguished by cartilage degradation and bone changes [30]. It is difficult to diagnose OA in its early stages because it has been suggested that the first changes occur in the subchondral bone [31] before the occurrence of joint space narrowing and osteophytes. As a result, early detection, diagnosis, and intervention are highly desired in order to overcome this severely disabling disease, as early-stage treatment could prevent cartilage and bone breakdown.

Thus, focusing on the implications of similarities and the potential scope of human errors being significant to the problem, there is a need for a strong model that can assist healthcare professionals in having more accurate diagnostics and an effective system in place for quick turnaround analysis with the appropriate type of detection for the type of arthritis. As a result, the focus of this chapter is on providing a detailed review of the current solutions or models used in ML and non-ML practices that could be useful in developing a more appropriate type of system [6].

The focus of the other sections of this chapter is on understanding the literature's possible set of solutions for using ML models for arthritis form detection.

## 4.2 LITERATURE REVIEW

This chapter's review of the literature is intensely focused on how computer-aided solutions are in place and the implications, if any, with the existing models, as well as the potential scope for the application of combinations to be used for improving the existing models or developing sustainable solutions that can improve the accuracy and adequacy in diagnosis.

The structure chosen for the literature review is to concentrate on the last seven years of research data in the form of peer-reviewed journal articles that can aid in the completion of the proposed research. The following are the objectives of the literature review, which served as a guide for the review.

### 4.2.1 RESEARCH OBJECTIVES

- To comprehend the current range of computer-aided solutions used in the diagnosis of RA and OA.
- The experimental studies' results successfully refer to the conditions of arthritis detection in order to understand the proposed ML models and the efficacy of the classifiers.
- Identify the symptoms or characteristics that are causing the problems and consider variables or metrics for more accurate detection.
- Thus, the review of research is discussed in detail in the following subsections, with a focus on the aforementioned objectives.

### 4.2.2 RELATED WORK

The authors of Ref. [7] concentrated on the dynamics of how ML models can be useful in identifying rheumatoid problems. The study discussed various potential solutions, such as supervised learning models and nonsupervised learning models, that could be useful for implementation. The study served as a comprehensive review of the possible applications, focusing on the dynamics of how various ML models and classifiers can assist. Another important aspect discussed in the study is the development of a systematic review, which aids in understanding the dynamics of how ML solutions can work as a systematic outcome for deciding on treatment conditions for patients based on data.

The authors of Ref. [8] discussed how ML models can be useful for the early detection of arthritis problems. Despite this, the study outlined the purpose of how the ML models can be viewed as an effective solution for arthritis detection. Most of the inputs discussed in the study are about the pros and cons of ML models, how the solutions can be resourceful for individuals, and the steps to be taken for the critical application of the ML models. The chapter review provided two key insights into various aspects of ML models, such as image-based detection and data-centric detection, structure recovery, significant data compression, and so on. Another important takeaway is a detailed review of references citing how ML models are used in other areas of diagnosis such as cardiology, oncology, endocrinology, and hematology. The study's findings can be used to consider how solutions can be pragmatic for improving the overall outcome of using ML for medical diagnostics.

Kataria and Ravindran [9] provide a detailed overview of how the e-health system can be a profound solution for dealing with business dynamics. Considering how various application systems, wearable devices, and tracking solutions can assist healthcare professionals in tracking the arthritis levels of various patients, it refers to pragmatic ways of using technology for better monitoring and treatment. The model also emphasizes the use of augmented data analysis patterns, which can aid in gaining insights into the health conditions of each individual patient, as well as the use of data in macroassessment to focus on the dynamics of data analysis for research and treatment care.

A research paper on digital health advocacy [10] discusses how various emerging technologies, such as wearable devices, pattern studies, big data analysis, real-time tracking sensors, and so on, can help improve the overall quality of healthcare management. The study emphasizes the importance of using sensor networks in the treatment of RA. These wearable devices can help track patient data, which can help improve clinical trials and conditions, resulting in better patient care. Despite the fact that the study is more of a comprehensive review model on digital health, the emphasis is on understanding how the technology scope improves the underlying conditions of patient data monitoring and related ways of applying the diagnosis models for detection accuracy.

The authors of the study [11] discussed the conditions of mathematical models and the treatment conditions significant for RA treatment, with a focus on quantitative predictive modeling applications for RA. The authors of the study emphasize the importance of identifying the problem of RA at an early stage in order to provide more effective treatment. The mathematical models discussed in the study demonstrate how statistical models such as the stochastic and progressive analysis models can be useful.

The study discussed ordinary differential equations (ODEs) and partial differential equations (PDEs) models, highlighting how the solutions are used in the RA progression analysis over two decades. The study emphasizes the absence of more focused studies on biological mechanisms for disease progression. The main point to consider from the condition above is to look into a possible solution in which the mathematical models can be adapted to the ML models. If this is the case, the number of ML models from the literature exploring the application of ODEs and PDE solutions to the detection of RA in patients will increase.

Sharon et al. [12] discuss the classification of medical disorders based on RA and orthopedics datasets using ensemble methods, where a modern model is proposed to determine RA using an ensemble model. A white blood cell classification was used to generate RA datasets. The models' feature extraction was based on lymphocytes obtained from a digital microscope using an electronic image sensor. Orthopedic datasets were used as the primary benchmark datasets for training the classifiers and testing the proposed model.

The proposed model incorporates three ensemble algorithm models: Adaboost, Bagging, and Random subspace. The KNN (*K*-nearest neighbor) and random forest (RF) algorithms were modified. In the study, data classification is modeled using 10-fold cross-validation and a holdout set-up.

According to the study, the accuracy ranges from 97.50% to 94.84% across a variety of ensemble classifiers. Bagging RF is advocated as an effective algorithmic model for the system's successful implementation. The study clearly shows that the ensemble classifier model is regarded as important in the management of arthritis conditions.

The studies in Ref. [13] focused on tertiary-care RA patients. The wrapper feature selection algorithm was used in the study to determine the attribute core set. Four distinct sets of algorithms were used to validate the model's dataset-based performance. The model's feature system focuses on DAS28 (Disease Activity Score on 28 joints) remission as the critical factor for analyzing the model. The ML models used in the model are KNN, extreme gradient boosting (XGBoost), RF, and logistic regression (LR).

Despite this, the model has focused on specific and niche feature segments. However, the study refers to the conditions in which the model was successful to the extent of 72.7% in the case of the XGBoost algorithm, and the other algorithms used for analysis have fetched a far lower level of accuracy. To summarize, the model is the most effective solution; however, the solution is less significant in terms of accuracy, which is critical in the model.

The study [14] discusses a magnetic resonance imaging (MRI) report-based imaging-centric ML model, with the authors emphasizing training the algorithms to perform significantly better than the proposed conditions. Understanding the conditions of bone marrow edema and tenosynovitis is the focus of the proposed features. The system's data study involves using 1,700 MR scans from hospital records as anonymous data. One of the model's critical implications is that, rather than adhering to any specific model in which the key inputs are used for analysis, the model encourages a qualitative understanding of the datasets. As a result, it is clear that the model under consideration requires more concrete validation of the precision, sensitivity, and specificity required to carry out the accurate analysis.

Another critical area where research has been conducted is disease activity prediction. The study authors considered using ML models to understand the level of RA problems faced by the patients in Ref. [15]. The study's significance lies in describing how the application of ML problems has exploded in RA segments. One of the key challenges discussed as an impacting factor in the model is understanding the gaps in the quality of data used for the analysis and the implications that could be encountered if the right kind of data is not used for training. In practical terms, the model is more qualitative and discusses how ML is a potential solution and the benefits of ML.

One of the most difficult challenges in RA treatment is how patients may experience a flare-up of the condition. Although the likelihood is low, it may have an adverse effect in cases where such issues are prevalent. Thus, in the study [16], the authors focused on training datasets derived from electronic medical records using various classifiers such as KNN, RF, LR, stacking meta classifier, and naive Bayes classifier. Kaggle datasets are used in the model's experimental analysis to ensure controlled implementation and understanding of clinical disease activity (DAS28 ESR). In order to keep the study more focused and exclusive, the impact in the study is limited to disease-modifying antirheumatic drug (DMARD) therapy. As a result, the study did not concentrate on synthetic DMARDs.

Lin et al. [17] have focused on using electronic medical records to assist users in gaining the right kind of insights into the early and automated prediction of potential RA issues in people. The study used medical records and the Apache Clinical Text Analysis and Knowledge Extraction System to gain the right insights into the data using the support vector machine (SVM) classifier. Based on the chi-square test and other statistical analyses performed on the model, the experimental study of the model discussed in the study indicates a moderate accuracy factor.

The authors of a novel study [18] attempted to assess how ML models can be resourceful for assessing RA using a series of datasets from the university hospital. The study, which used 820 patient records, discusses how, with the right kind of training, the model's accuracy can be improved. The prior Crohn's Disease Activity Index score, erythrocyte sedimentation rate (ESR), C-reactive protein (CRP) level, and DMARDs are the key variables used in the ML analysis in the models. Based on the preliminary description, it is concluded that the application of ML solutions in RA detection is feasible.

A recent study [19], focusing on the efficacy of ML models in comparison to the traditional approach of MTX (methotrexate), refers to a distinct set of features that can be useful for effective prediction. The logistic regression factor was used to analyze the models using the DAS28. The model's key classifiers are XGBoost, Least Absolute Shrinkage and Selection Operator (selection operator), and RF models. The model's experimental study results are evaluated using the performance of various parameters. According to the study, the logistic regression approach performs better at an 81% sensitivity rate, whereas the accuracy of MTX is around 69% when similar features are assessed. As a result, the study advocates the use of six critical features in ML.

In a study [20] conducted in Korea, the authors correlated aging and rheumatism by analyzing four key attributes for assessing RA conditions among patients: RA factor anticyclic citrullinated peptide, swollen joint count, and ESR. The proposed solution employs a model that employs *K*-means to study the threshold values of the four parameters on a regular basis. The approach used in the study is not limited to predicting RA conditions in a patient at the early stages but also takes the progressive stages into account. Based on the positive values of two of the four factors, RF and AC, the study was able to estimate the occurrence of rheumatoid disease. When RF and AC are used as metrics for assessment, experimental studies related to the discussion on the model refer to improved prediction accuracy. As a result, the ML approach can be used to detect rheumatic disease and improve the quality of life in an aging society.

A recent study [21] focused on the application of ML models to OA, which is a more common type of arthritis problem that occurs in the knees. The model discussed the applicability of two methods for predicting knee joint space narrowing (knee-JSN). The model discussed in the study involves the use of the clustering process to identify groups of people who are at different stages of nonprogressing and progressing JSN. A robust set of feature selection approaches, including embedded techniques, a wrapper, and a filter, allows for the evaluation of risk factors used in the model. The experimental study's approach to assessing both single and both knee-JSN progression refers to a higher accuracy rate of 83.3% when the instances assess both knees, rather than individual knee condition assessment.

According to the authors of the study [22], a novel approach of patient-specific analysis and prediction model for OA is being considered. The study examines existing models in order to identify gaps in the early stages of OA disease detection. It emphasizes the categorization of various classifier-centric studies, as well as their accuracy conditions. A review of the summation study led to more detailed and implicit studies that discussed a large set of ML models for OA analysis.

Kokkotis et al. [23] discuss an evolutionary ML approach in which the study targeted the risk factors for KOA progression using a significant set of feature selection methodologies. The model addresses two major challenges: issues of dimensionality and heterogeneity for the data obtained from the database, and secondarily, imbalance issues observed in the KOA progressors class in comparison to the nonprogressors class.

The study has primarily relied on the Gen wrapper feature selection model for applying the features for accuracy. The study used a genetic algorithm model to improve the model's accuracy when it came to evolutionary models. The Shapley Additive explanation is used to evaluate the datasets used for the filtered feature selection efficacy. The proposed model is promoted as an effective risk stratification strategy that, when applied to KOA patient data prediction, can aid in more accurate analysis.

A recent study on KOA prediction focused on the patient's need for total knee arthroplasty. The domain's predictive model refers to KOA conditions in the general population, issues of TKA outcome in KOA patients, and the risk of knee replacement conditions in KOA patients. Using the PubMed databases, certain models from the literature were discussed, emphasizing the significance of feature selection and conditional approaches from algorithms that can help assess KOA conditions among patients more effectively. The model's key aspects are prognostic prediction models that are qualitatively summarized. The study used a meta-analysis approach to identify quantitative metrics for the predictive performance of the models.

Another study focusing on the conditions of early warning signs in the case of OA structural progression proposes a new solution based on

ML that focuses on detecting the progressions at early stages. The study focused on feature selections based on the bone mass index by using baseline serum levels of six adipokines and their respective ratios and inflammatory factors. The model's key data are X-rays and three magnetic resonance imaging variables that can be used more effectively for effective prediction. The ML feature classification methodology is combined with five different ML algorithms. The experimental studies conducted by the study claim SVM with better accuracy conditions in the case of gender-centric classifier development [24]. According to the study, the proposed model application has more than 92% accuracy, as well as early-stage detection accuracy.

Other important models [22,24–29] discussed the conditions for applying the OA-related ML models. According to the comprehensive review of the studies, there is a diverse range of features used for the analysis of OA and RA. Prediction needs to be more accurate in order to improve solutions that can aid in early-stage diagnosis.

## 4.3 CHALLENGES

- Other body sounds, such as muscle activity, were difficult to distinguish from articular cartilage sounds. Knee radiographic images are extremely sensitive to unintended defects, which can complicate the study of bone structures. It is possible that the experts will take more time to analyze the knee X-ray and determine the presence of OA.
- Hand-drawn contours and joint area cropping are used for the region of interest segmentation, which is tedious and time-consuming.
- The primary cause of geometric distortions of the cartilage region in knee X-ray images is the progression of OA, which may be misrepresented due to image acquisition filming, handling, and digitization. Extraction of significant regions from such distorted images may become difficult.
- Extraction of significant regions from such distorted images may become difficult.
- It is also difficult to recover the relevant region when images are distorted due to geometric deformation.

- Current imaging methods, such as X-ray, MRI, and ultrasound, have low sensitivity in early disease, and as a result, OA is already advanced at the time of diagnosis, and understanding of its cause and progression is still limited.

## 4.4 PROBLEM STATEMENT

OA is one of the most common chronic joint diseases, with a large number of risk factors that are directly linked to it. When compared to other types of OA, KOA is the most common and occurs most frequently in people over the age of 55, with the disease's prevalence increasing as people reach the age of 65. KOA is difficult to diagnose in the early stages of the disease. To address this issue, we are developing an ensemble approach early detection deep learning-based system that will use an X-ray/MRI dataset of KOA patients to provide more accurate results than previous work.

## 4.5 PROPOSED METHOD

The prevalence of OA is increasing, which is partly due to our aging population. OA is currently diagnosed using patient-reported symptoms and X-rays. The alternative is an MRI, but this comes at a high cost and is rarely used until symptoms worsen and patients are referred for a surgical opinion from a specialist. As a result, effective management and early detection of KOA is a critical public health issue that concerns the general public as well as a variety of clinicians and healthcare executives.

Many researchers have attempted to diagnose OA using X-rays and patient medical data, but the results have been inconsistent. Through this proposed idea/work, we will design and implement an ensemble methodology for early detection of KOA using a hybrid dataset of X-ray and medical data of patients through which doctors can predict the OA and its level more accurately and in less time through this proposed idea/work. Image processing and AI will be used to create a low-cost methodology. Figure 4.1 depicts the proposed system architecture.

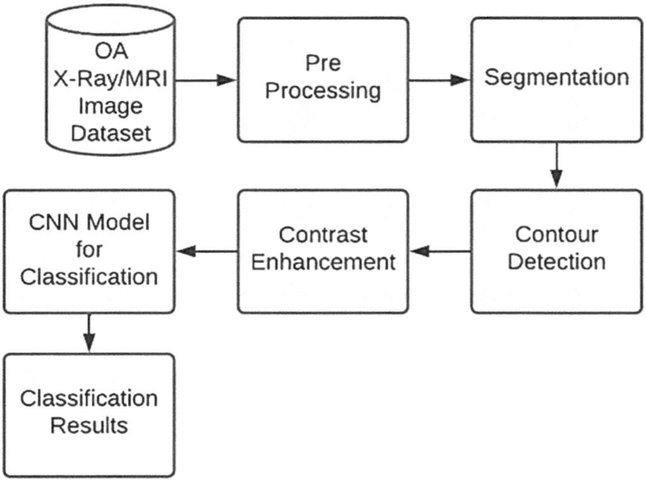


**FIGURE 4.1** Architecture of the proposed system.

## 4.6 PROPOSED SYSTEM AND ITS ADVANTAGES

In the base article or earlier work referred to (A New Approach to Classify Knee Osteoarthritis the Severity from Radiographic Images Based on CNN LSTM Method by Rima Tri Wahyuningrum & Other Co-Authors), the author had used X-ray image dataset and implemented CNN model, but the accuracy was obtained as 75.28%, whereas in our proposed system we will be using hybrid dataset including X-ray/MRI and with advance image preprocessing and CNN model, we will be achieving higher accuracy which will help to predict the KOA with the more accurate result at an early stage.

**KEYWORDS**

- **machine learning**
- **rheumatoid arthritis**
- **osteoarthritis**
- **artificial intelligence**
- **magnetic resonance imaging**

## REFERENCES


1. https://www.uofmhealth.org/conditions-treatments/cmc/difference-between-osteoarthritis-and-rheumatoid-arthritis#:~:text=Osteoarthritis%20occurs%20when%20the%20smooth,the%20body%20instead%20of%20intruders.
2. Wardito E, Purba HH, Purba A. System dynamic modeling of risk management in construction projects: A systematic literature review. *Operational Research in Engineering Sciences: Theory and Applications* 2021;4(1):1–8.
3. Nelson AE, Fang F, Arbeeva L, Cleveland RJ, Schwartz TA, Callahan LF, Marron JS, Loeser RF. A machine learning approach to knee osteoarthritis phenotyping: Data from the FNIH biomarkers consortium. *Osteoarthritis and Cartilage* 2019;27(7):994–1001.
4. Maarseveen TD, Meinderink T, Reinders MJ, Knitza J, Huizinga TW, Kleyer A, Simon D, van den Akker EB, Knevel R. Machine learning electronic health record identification of patients with rheumatoid arthritis: Algorithm pipeline development and validation study. *JMIR Medical Informatics* 2020;8(11):e23930.
5. Mate GS, Patil MS, Chavan MS, Harshada MS. Automatic prediction of rheumatoid arthritis using CNN. *Journal of Architecture & Technology* 2019;XI:18–23.
6. Stafford IS, Kellermann M, Mossotto E, Beattie RM, MacArthur BD, Ennis S. A systematic review of the applications of artificial intelligence and machine learning in autoimmune diseases. *NPJ Digital Medicine* 2020;3(1):1–1.
7. Hügle M, Omoumi P, van Laar JM, Boedecker J, Hügle T. Applied machine learning and artificial intelligence in rheumatology *Rheumatology Advances in Practice* 2020;4(1):rkaa005.
8. Kim KJ, Tagkopoulos I. Application of machine learning in rheumatic disease research. *The Korean Journal of Internal Medicine* 2019;34(4):708.
9. Kataria S, Ravindran V. Emerging role of eHealth in the identification of very early inflammatory rheumatic diseases. *Best Practice & Research Clinical Rheumatology* 2019;33(4):101429.
10. Kataria S, Ravindran V. Digital health: A new dimension in rheumatology patient care. *Rheumatology International* 2018;38(11):1949–1957.
11. Macfarlane FR, Chaplain MA, Eftimie R. Quantitative predictive modelling approaches to understanding rheumatoid arthritis: A brief review. *Cells* 2020;9(1):74.
12. Sharon H, Elamvazuthi I, Lu CK, Parasuraman S, Natarajan E. Development of rheumatoid arthritis classification from electronic image sensor using ensemble method. *Sensors* 2020;20(1):167.
13. Venerito V, Angelini O, Fornaro M, Cacciapaglia F, Lopalco G, Iannone F. A Machine learning approach for predicting sustained remission in rheumatoid arthritis patients on biologic agents. *JCR: Journal of Clinical Rheumatology* 2021;28(2):e334–e339.
14. Stoel BC. Artificial intelligence in detecting early RA. *Seminars in Arthritis and Rheumatism* 2019;49(3):S25–S28.
15. https://rheumatology.medicinematters.com/artificial-intelligence/devices-and-technology/machine-learning-treatment-rheumatic-disease/17540860
16. Vodencarevic A, Tascilar K, Hartmann F, Reiser M, Hueber AJ, Haschka J, Bayat S, Meinderink T, Knitza J, Mendez L, Hagen M. Advanced machine learning for

predicting individual risk of flares in rheumatoid arthritis patients tapering biologic drugs. *Arthritis Research & Therapy* 2021;23(1):1–8.
17. Lin C, Karlson EW, Canhao H, Miller TA, Dligach D, Chen PJ, Perez RN, Shen Y, Weinblatt ME, Shadick NA, Plenge RM. Automatic prediction of rheumatoid arthritis disease activity from the electronic medical records. *PLoS One* 2013;8(8):e69932.
18. Norgeot B, Glicksberg BS, Trupin L, Lituiev D, Gianfrancesco M, Oskotsky B, Schmajuk G, Yazdany J, Butte AJ. Assessment of a deep learning model based on electronic health record data to forecast clinical outcomes in patients with rheumatoid arthritis. *JAMA Network Open* 2019;2(3):e190606–e190606.
19. Gosselt HR, Verhoeven M, Bulatović-Ćalasan M, Welsing PM, de Rotte MC, Hazes JM, Lafeber FP, Hoogendoorn M, de Jonge R. Complex machine-learning algorithms and multivariable logistic regression on par in the prediction of insufficient clinical response to methotrexate in rheumatoid arthritis. *Journal of Personalized Medicine* 2021;11(1):44.
20. Yoo J, Lim MK, Ihm C, Choi ES, Kang MS. A study on prediction of rheumatoid arthritis using machine learning. *International Journal of Applied Engineering Research* 2017;12(20):9858–9862.
21. Ntakolia C, Kokkotis C, Moustakidis S, Tsaopoulos D. Prediction of joint space narrowing progression in knee osteoarthritis patients. *Diagnostics* 2021;11(2):285.
22. Jamshidi A, Pelletier JP, Martel-Pelletier J. Machine-learning-based patient-specific prediction models for knee osteoarthritis. *Nature Reviews Rheumatology* 2019;15(1):49–60.
23. Kokkotis C, Moustakidis S, Baltzopoulos V, Giakas G, Tsaopoulos D. Identifying robust risk factors for knee osteoarthritis progression: An evolutionary machine learning approach. In: *Healthcare*. Multidisciplinary Digital Publishing Institute, 2021, Vol. 9, No. 3, p. 260.
24. Bonakdari H, Jamshidi A, Pelletier JP, Abram F, Tardif G, Martel-Pelletier J. A warning machine learning algorithm for early knee osteoarthritis structural progressor patient screening. *Therapeutic Advances in Musculoskeletal Disease*. 2021;13:1759720X21993254.
25. Lazzarini N, Runhaar J, Bay-Jensen AC, Thudium CS, Bierma-Zeinstra SM, Henrotin Y, Bacardit J. A machine learning approach for the identification of new biomarkers for knee osteoarthritis development in overweight and obese women. *Osteoarthritis and Cartilage* 2017;25(12):2014–2021.
26. Widera P, Welsing PM, Ladel C, Loughlin J, Lafeber FP, Dop FP, Larkin J, Weinans H, Mobasheri A, Bacardit J. Multi-classifier prediction of knee osteoarthritis progression from incomplete imbalanced longitudinal data. *Scientific Reports* 2020;10(1):1–5.
27. Kokkotis C, Moustakidis S, Papageorgiou E, Giakas G, Tsaopoulos DE. Machine learning in knee osteoarthritis: A review. *Osteoarthritis and Cartilage Open* 2020;100069.
28. Tiulpin A, Klein S, Bierma-Zeinstra SM, Thevenot J, Rahtu E, van Meurs J, Oei EH, Saarakkala S. Multimodal machine learning-based knee osteoarthritis progression prediction from plain radiographs and clinical data. *Scientific Reports* 2019;9(1):1–11.
29. Jamshidi A, Leclercq M, Pelletier JP, Labbe A, Abram F, Droit A, Martel-Pelletier J. Developing a comprehensive patient-specific disease progression prediction

model for knee osteoarthritis using machine/deep learning methods. *Arthritis & Rheumatology* 2019;71:111.
30. Goldring MB, Goldring SR. Articular cartilage and subchondral bone in the pathogenesis of osteoarthritis. *Annals of the New York Academy of Sciences* 2010;1192(1):230–237.
31. Radin EL, Paul IL, Tolkoff MJ. Subchondral bone changes in patients with early degenerative joint disease. *Arthritis & Rheumatism*. 1970;13(4):400–405.

# CHAPTER 5

# Performance Evaluation of Heart Disease Classification Using Deep and Machine Learning-Based Methods

ADYASHA RATH[1*] and GANAPATI PANDA[2]

[1]*Department of Computer Science and Engineering, C.V. Raman Global University, Bhubaneswar, Odisha, India*

[2]*Department of Electronics and Tele Communication, C.V. Raman Global University, Bhubaneswar, Odisha, India*

*Corresponding author

## ABSTRACT

This chapter provides a simplified overview of early detection of coronary heart disease (CHD) from different types of data sources of subjects. It also provides separate block diagram with concise explanation of machine learning (ML)/deep learning (DL) classification models with different types of inputs. It deals with the development and performance evaluation of naïve Bayes (NB), long-short term memory (LSTM), generative adversarial network (GAN), and ensemble model of GAN and LSTM. The performance metrics in terms of accuracy, F1-score and area under the curve (AUC) obtained by simulating DL/ML/ensemble models along with receiver operating characteristics (ROC) plots are also provided in this chapter. It is observed from the results of the simulation study that the GAN model provides the best performance in terms of accuracy (0.978), F1-score (0.975), and AUC value (0.981) for MIT-BIH ECG dataset compared to LSTM and NB models. Further, it is observed that the ensemble model developed using GAN and LSTM methods yield the

Machine Learning Applications for Data Analysis in Healthcare Systems: Sudeshna Chakraborty, Jyotsna Singh, Praveen Kumar Shukla, Prasenjit Chatterjee (Eds.)


DOI: 10.1201/9781779643193-5

overall best performance having accuracy, F1-score and AUC values of 0.992, 0.987 and 0.984, respectively. It is expected that this chapter will provide a good exposure to the new researchers in this field.

## 5.1 INTRODUCTION

Cardiovascular disease (CVD), commonly known as heart disease (HD), is the main cause of global mortality. One-third of the total death in HD occurs below 70 years of age. The main causes of CVD are due to smoking, consumption of alcohol, obesity, overweight, high blood pressure, and abnormal glucose level. CVD mainly refers to disorders of heart and blood vessels. As per the World Health Organization, the number of deaths will substantially increase in the next 10 years.

This chapter presents a complete overview of types of HDs and the associated causes, the types of test features, the types of signals source from the body, the various common features extracted for detection purpose. In addition, the common methodology used for early detection of CVD as well as the various machine learning (ML) and deep learning (DL) methods employed are also presented. The standard databases which are mostly being used by HD research community have also been listed for convenience of carrying out further research in this field. The generalized training and validation schemes commonly used for the development of HD detection have been outlined. Some of the key results and findings reported in recent literature are also presented to have a complete understanding on CVD detection. It is expected that the materials that are presented subsequently in this chapter would be immensely useful for carrying out further research as well as for development of Internet of Things (IoT) and cloud computing-based HD diagnosis system.

## 5.2 OBJECTIVE AND ORGANIZATION

The main objectives as well as the organization details of this chapter are presented.

### *5.2.1 OBJECTIVES*

1. To provide an overview of the types of CVD and brief explanation on each type of CVD. To present various standard datasets which are universally used by the researchers as well as a list of various input signal and test data extracted from the patient for the diagnosis of CVD. To outline a list of standard feature extraction techniques that are usually employed for better diagnosis and for reduction of computational complexity.
2. To develop ML, DL, and IoT based as well as ensemble models for detection of CVD methods, respectively.
3. To present various ML and DL methods which are commonly being used for classifying normal and abnormal patients.
4. To provide sample simulation-based results and few performance metrics to assess the efficacy of detection as well as to highlight the contribution of this chapter.
5. To mention future scope of research work on CVD detection.

### *5.2.2 ORGANIZATION*

Section 5.3 deals with types of CVD, standard datasets, and various types of clinical tests and signals that are used for CVD detection. The various methods for features extraction and reduction as well as different commonly used ML and DL methods are also presented in Section 5.4. In Section 5.5, simplified schemes of ML, DL, and IoT-based CVD classification have been explained. It also presents a generalized ensemble model for CVD detection. Some simulation results in terms of plots, performance measures obtained from the experimental study are presented in Section 5.6. The contribution of this chapter is also outlined in the same section. Finally, the conclusion and future scope of the extension of the work are dealt in Section 5.7.

## 5.3 TYPES OF CVD, CLINICAL TEST, SIGNALS, AND STANDARD DATASETS

This section lists and explains the various types of CVDs that usually human beings are attacked. In addition, the various clinical test information as well as types of signals extracted for the detection purpose has been presented. To facilitate further research the standard datasets have been documented. There are nine types of HDs which are presented in Table 5.1. It also explains briefly on each type of CVD.

**TABLE 5.1** Types of CVD

| S. No. | Types | Definition/Explanation No. |
|---|---|---|
| i. | VHD/HVD | The VHD/HVD comprises of stenosis, prolapse, and regurgitation. The stenosis occurs due to narrow opening of valves. Improper closing of valves leads to prolapse. The backward leaking of blood causes regurgitation |
| ii. | Pulmonary embolism | The blockage in one of the pulmonary arteries in the lungs leads to pulmonary embolism |
| iii. | CAD/CHD/IHD | The narrowing of heart arteries (supply blood to the heart muscle) is the cause of IHD/CHD/CAD |
| iv. | Congenital heart disease | This type of HD occurs due to septal defect hole between two heart chambers. It also occurs due to obstruction defect which blocks free flow of blood. The Cyanotic HD is due to shortage of oxygen. This congenital HD is present since birth |
| v. | Cerebrovascular disease | This type of HD affects the blood vessel as well as the supply of blood to the brain |
| vi. | Peripheral heart disease | If the blood vessel supplying to the arms and legs are affected, it gives rise to peripheral HD |
| vii. | Cardiomyopathies | If the muscle of the heart is affected, it leads to cardiomyopathies |
| viii. | Rheumatic heart disease | The streptococcal bacteria damage the heart muscle and valves. This gives rise to rheumatic HD |
| ix. | HF | The HF occurs if the heart does not pump blood properly around the body. It may be due to the effect of left and right side of the heart |

*Notes*: VHD: Valvular heart disease/HVD: heart valve disease, CAD: coronary artery disease, CHD: coronary heart disease, IHD: ischemic heart disease, HD: heart disease, HF: heart failure.

### *5.3.1 TYPES OF INPUTS USED FOR CVD DETECTION*

For classification of CVD, various inputs from the subjects are fed separately to detect abnormality in the functioning of the heart. The features from these inputs are obtained by different techniques and subsequently these are used to the appropriate classifiers to detect the CVD. These are listed and explained in Table 5.2.

**TABLE 5.2** Types of Inputs Used for CVD Detection

| Types | Explanation |
|---|---|
| Clinical test | Chest pain, resting blood pressure, serum cholesterol, fasting blood sugar, resting electrocardiographic, maximum heart rate, exercise-induced angina |
| ECG | ECG is a graphical plot showing the interrelation between the electrical activity of the heart versus time. It happens due to depolarization and repolarization of cardiac muscle |
| Heart sound | The heart sound is created due to turbulence when the heart valves snap shut |
| ICG | It is noninvasive in nature and measures the total electrical conductivity of the thorax. It changes in time to process continuously a number of cardiodynamic parameters |
| MCG | It is a technique to measure the magnetic field strength produced by electrical current in the heart |
| CT images | It is a noninvasive imaging test of the arteries that supply blood to the heart. It is used to diagnose the condition of the heart |

*Notes*: ECG: electrocardiogram, ICG: impedance cardiography, MCG: magnetocardiography, CT: computed tomography

## 5.3.2 SOURCES OF DATA FOR CVD DETECTION

The CVD detection models are developed through training process using standard data relating to CVD. Subsequently, these models are applied to diagnose the HD of a patient. These standard datasets have been collected and presented in Table 5.3 [1]. In many situations, real-life data are collected from various standard hospital sources preprocessed and passed through intelligent classifiers for detection. In addition, online medical data are obtained from patients at remote places and then transferred to doctors in standard hospitals through IoT platform.

**TABLE 5.3** Details of Standard CVD Datasets [1]

1. UCI ML Repository
    i. Cleveland HD dataset
    ii. Hungarian HD dataset
    iii. SPECTF cardiac dataset
    iv. StatLog HD dataset
    v. Switzerland dataset
    vi. Heart disease dataset
    vii. SPECT HD dataset
    viii. Pima Indian Diabetes dataset
2. STULONG dataset
3. MITDB dataset (PhysioNet)
4. MITBIH arrhythmia dataset
5. St. Petersburg Institute of Cardiological Arrhythmia dataset
6. Z-Alizadeh Sani dataset
7. FuWai dataset
8. PhysioNet dataset
9. PTB-ECG dataset

## 5.4 FEATURES EXTRACTION, REDUCTION, AND CLASSIFICATION TECHNIQUES

For efficient classification between normal and abnormal as well as to train the model faster, features/reduced features are fed. These features are obtained from either ECG, heart sound, speech, ICG, MCG, or CT images. The various methods of feature extraction are as follows: discrete wavelet transform (DWT) [2], Mel-frequency cepstral coefficients (MFCC) [3, 4], entropy [5], and artificial neural network (ANN) [6, 7]. If the number of features is large, feature reduction techniques such as Linear Discriminant Analysis (LDA) [8], Ruzzo–Tompa approach [9], and principal component analysis (PCA) [10, 11] are used for reducing the number of features. In some situations, features are selected based on contribution to classification accuracy in case of using bio-inspired techniques. Some standard features extraction and reduction methods which are used CVD detection have been listed in Table 5.4.

**TABLE 5.4** Methods of Features Extraction/Selection and Features Reduction

| Features Selection/Features Extraction | Features Reduction |
|---|---|
| DWT [2] | LDA [8] |
| MFCC [3] | Ruzzo-Tompa approach [9] |
| Entropy [5] | Bio-inspired techniques |
| ANN [6] | PCA [10] |

The classification task is very vital for detecting abnormal heart conditions. The conventional classifiers used belong to both ML- and DL-based approaches. Table 5.5 presents a few standard and recently used ML and DL techniques for CVD detection. The left column of Table 5.5 shows few standard methods of ML techniques whereas the right column indicates some DL-based classifiers for CVD.

**TABLE 5.5** ML and DL Techniques Used in CVD Detection

| ML and DL Techniques | |
|---|---|
| ANN [12] | ELM [13] |
| RF [14] | CRNN [15] |
| NB [12] | RNN [16] |
| DT [12] | CNN [17, 18] |
| SVM [12] | LSTM [19] |
| LR [12] | DBNN [20] |
| HMM [3] | GAN [21] |
| Boosted tree [22] | RBM [5] |
| AdaBoost [22] | Deep ANN [23] |

*Notes*: ANN: artificial neural network, ELM: extreme learning machine, RF: Random forest, CRNN: convolutional recurrent neural network, NB: naïve Bayes, RNN: recurrent neural network, DT: decision tree, CNN: convolutional neural network, SVM: support vector machine, LSTM: long short-term memory, LR: logistic regression, DBNN: deep belief neural network; HMM: hidden Markov model; GAN: generative adversarial network, RBM: restricted Boltzmann machine.

## 5.5 GENERALIZED ML, DL, AND IOT PLATFORM-BASED CLASSIFIER MODELS FOR CVD DETECTION

### *5.5.1 ML-BASED CVD DETECTOR*

In this section, the generalized block diagrams of various classification schemes have been presented. Figure 5.1 provides an adaptive classification model for detecting CVD. This scheme takes samples of either ECG signal/heart sound/clinical test data/ICG/MCG/CT images as inputs. Subsequently features are extracted and if needed these are reduced to make the classifier simple. These features are then fed to the model for detection purposes. The list of features extraction methods is given in Table 5.4. The extracted HD-related features are passed through any of the classifiers listed in Table 5.5. In supervised learning approach, the error term computed between desired and predicted classes are used in the learning algorithm to update the trained parameters of the classifier. After successful training of the classifiers, the parameters are frozen. The trained model is subsequently used for validation purpose using the feature sets of patients which have not been used for training purpose.

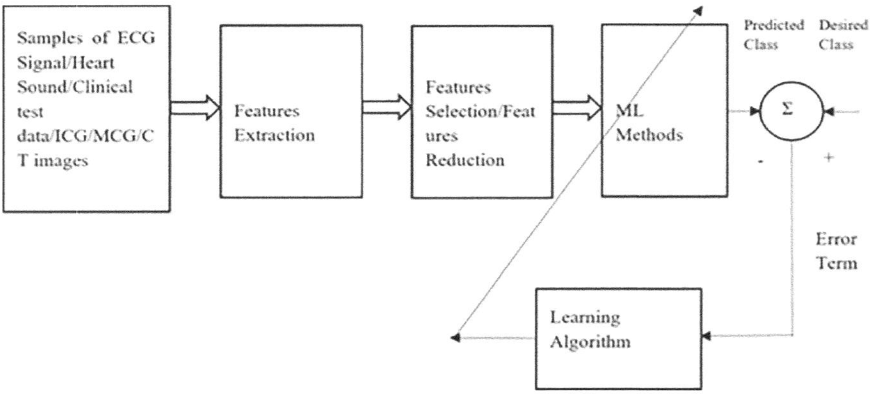


**FIGURE 5.1** Generalized block diagram of ML-based CVD detection model.

### *5.5.2 THE DL-BASED CVD DETECTOR*

Obtaining the desired features for achieving better classification is a complex task. If the features chosen are not appropriate, the classification accuracy goes down. To alleviate this problem, the DL-based classifiers have been reported in the literature which essentially do two main tasks: features extraction and classification. The block diagram of generalized DL-based classifier is shown in Figure 5.2. As may be evident, one single block contains both features extraction and classifier. For example, in the case of CNN classifier, the set of convolution operations extract the desired features. The pooling operation reduces the size of features vector.

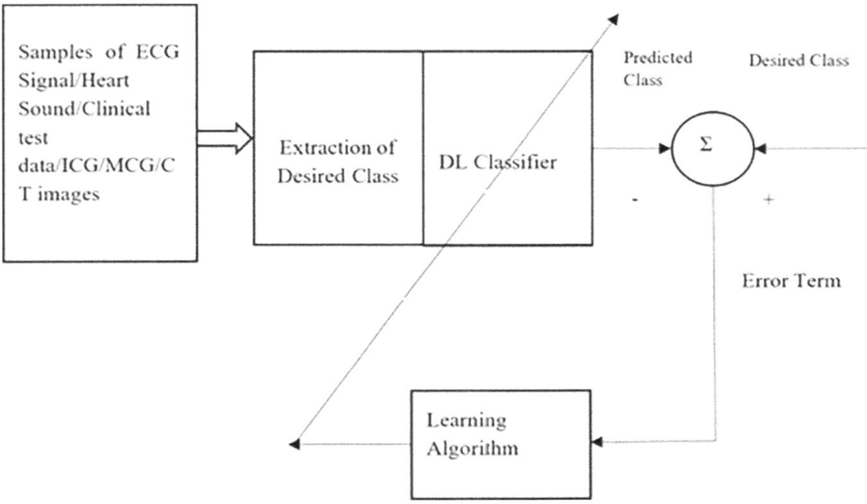


**FIGURE 5.2** Generalized block diagram of DL-based CVD detection model.

### *5.5.3 IOT PLATFORM-BASED CLASSIFIER FOR CVD DETECTION*

The IoT platform-based CVD detection model is shown in Figure 5.3. The first block contains the direct noninvasive test signal obtained from the patients of remote places. This extracted signal (ECG/sound signal/ICG/MCG/CT images) is discretized and the samples are passed to cloud data center through sensor network and gateway. In this case also, the samples are passed through the preprocessing block comprising of low pass filter followed by wavelet transform-based denoiser. The preprocessed data is then used to extract features extraction and subsequently used for the classification of CVD.

### *5.5.4 ENSEMBLE CLASSIFIER FOR CVD DETECTION*

Conventionally, the accuracy achieved by a single ML or DL method may not be high when some standard datasets or imbalanced datasets are used. For such situation, if the CVD classification accuracy is

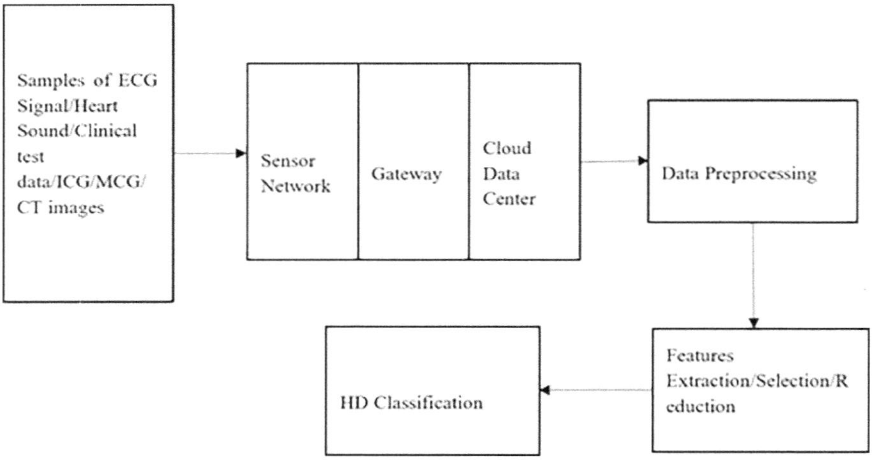


**FIGURE 5.3** Block diagram of IoT-based CVD detection model.

to be improved, then ensemble model needs to be developed and used by suitably combining the predicted output of two or three classifiers. In Figure 5.4, the generalized block diagram of an ensemble CVD detection model is shown. At first, two ML/DL models are trained individually using features of standard HD datasets. These trained ML/DL models are placed in parallel and the predicted outputs of each individual model are weighted (by $W_1$ and $W_2$) as shown in Figure 5.4. There are many methods by which these weights are to be optimized. Some common methods used for developing of an ensemble

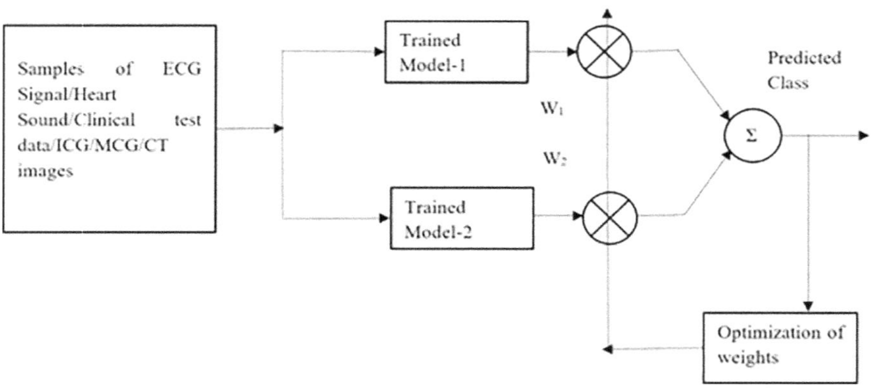


**FIGURE 5.4** Block diagram of ensemble classification model for CVD detection.

model are majority voting scheme and bio-inspired-based weight optimization scheme. In most cases, the ensemble model outperforms the individual models for CVD detection.

## 5.6 EXPERIMENTAL RESULTS

The performance measures during training and validation phases have been evaluated by simulating the naïve Bayes (NB), long short-term memory (LSTM), and generative adversarial network (GAN) models using MIT-BIH HD dataset. In addition, the ensemble model of LSTM-GAN is also simulated using the same dataset. The various results have been obtained and the details have been presented in Ref. [18]. Some of the reported results of this article have been provided in this section for ease of understanding and to carry out further research in this field.

The comparison of accuracy plots of NB-based CVD detection model during training and validation phases is shown in Figure 5.5. It is observed, as expected, that the accuracy during the validation phase is lower than that during training phase. The accuracy increases steadily during training until it achieves almost a constant value after successful training of the model. A similar plot for LSTM model is presented in Figure 5.6. In the case of GAN-based DL model similar accuracy plot has been obtained through a simulation study and is shown in Figure 5.7. This plot exhibits the comparison of accuracy of generator and discriminaror parts of GAN. The receiver operating characteristics (ROC) plot of a model is very important as its area under the curve (AUC) signifies the performance potentiality of a detector. The ROC plots of various ML and DL models have been obtained using simulation study and by MIT-BIH dataset. The comparative ROC plots of various classification models are shown in Figure 5.8. Also the corresponding AUC value of each model is listed in the same figure. It may be noted that the higher the AUC, the better is the detection efficiency of a model. The three performance measures: Accuracy, $F_1$ score, and AUC of NB, LSTM, GAN, and ensemble of GAN-LSTM models obtained from the simulation study and have been presented in Table 5.6. It is observed from Table 5.6 that GAN-LSTM ensemble model provides the best performance of classification yielding 0.992, 0.987, and 0.984 values in terms of accuracy, $F_1$ score, and AUC, respectively. In descending order of these performances, the second, third, and fourth ranks of the models are GAN, LSTM, and NB. To assess the consistent performance of these models, simulation study using other datasets should also be carried out and the results need to be analyzed.

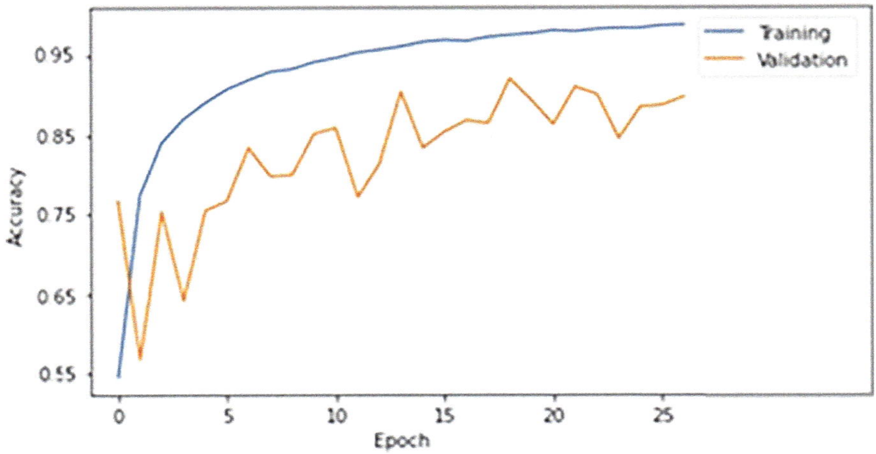


**FIGURE 5.5** Comparison of accuracy plots of NB model for MIT-BIH dataset [18].

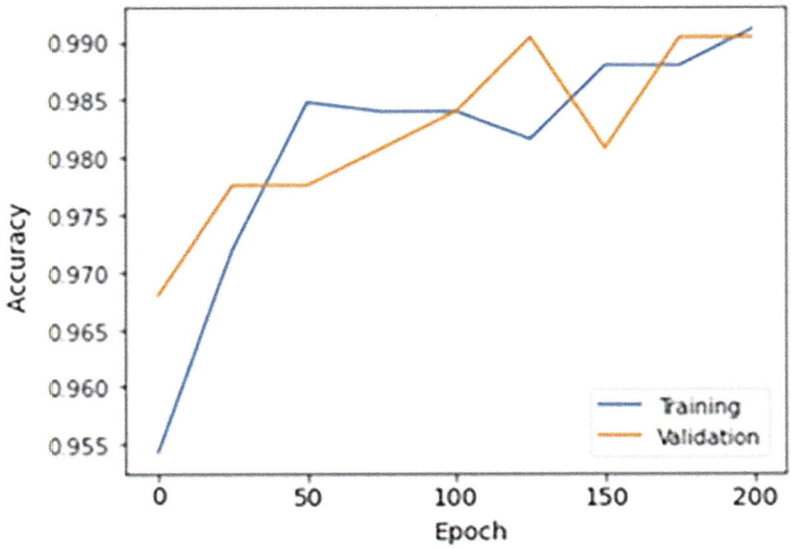


**FIGURE 5.6** Comparison of accuracy plots of LSTM model for MIT-BIH dataset [18].

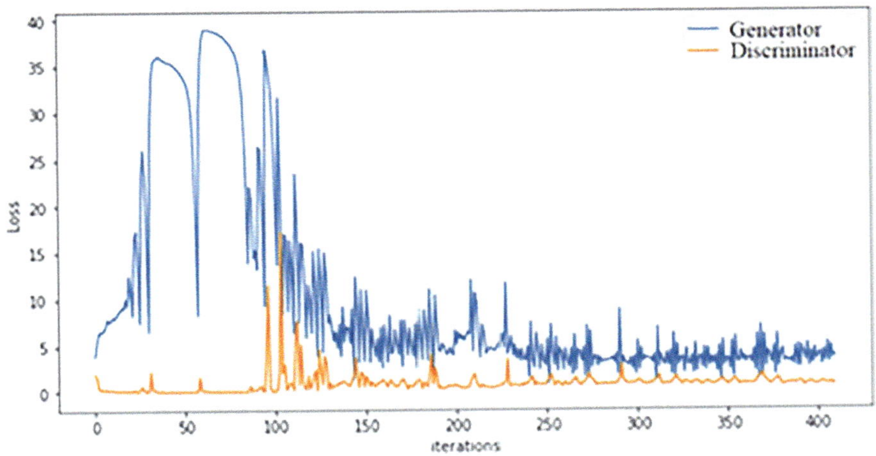


**FIGURE 5.7** Comparison of accuracy plots of GAN model for MIT-BIH dataset [18].

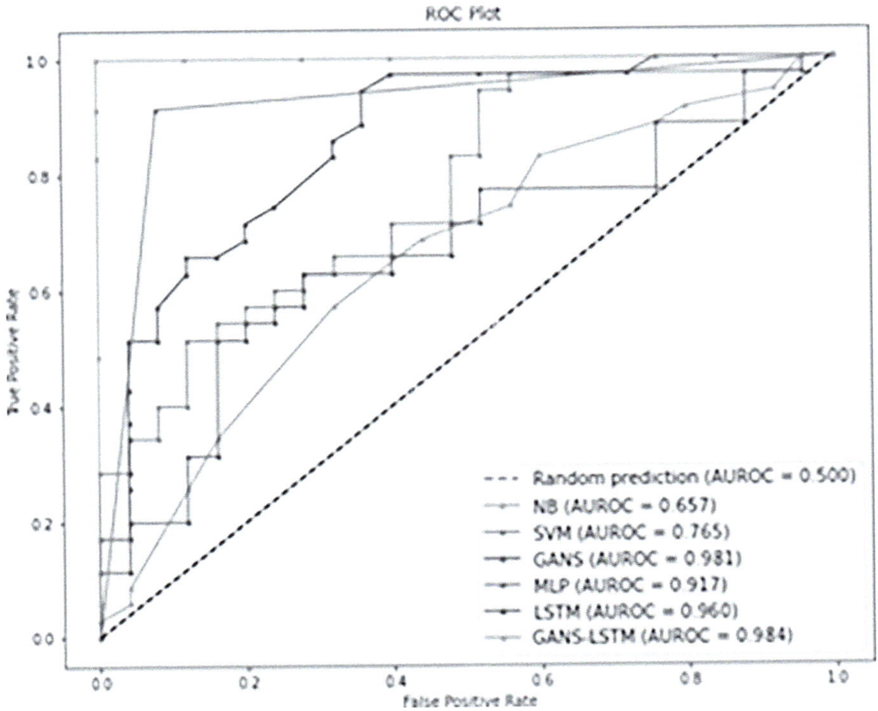


**FIGURE 5.8** Comparison of ROC Plots of different models using MIT-BIH dataset [18].

**TABLE 5.6** Comparison of Three Performance Measures of CVD Detection Using MIT-BIH Dataset [18]

| Performance Measures | NB | LSTM | GAN | GAN-LSTM |
|---|---|---|---|---|
| Accuracy | 0.821 | 0.966 | 0.978 | 0.992 |
| $F_1$ score | 0.817 | 0.937 | 0.975 | 0.987 |
| AUC | 0.657 | 0.960 | 0.981 | 0.984 |
| Best performing model | GAN-LSTM | | | |

## 5.7 CONCLUSION AND FUTURE SCOPE

An overview of ML, DL, and ensemble models for early detection of CVD has been presented. This will help for early diagnosis and thus many human lives can be saved. By analyzing the presented simulation results it is demonstrated that the GAN–LSTM ensemble model outperforms individual DL model as well as ML models. The presentation of the text in this chapter has been made in a simplified way to facilitate interested researchers and learners to carry out research work in this important and emerging field. To learn more, details on various papers are listed in the reference.

## KEYWORDS

- **detection of HD using ML**
- **IoT-based HD**
- **ECG samples-based HD detection**
- **ICG-based HD detection**

## REFERENCES

1. Rath A, Mishra D, Panda G et al. An exhaustive review of machine and deep learning based diagnosis of heart diseases. *Multimedia Tools and Applications* 2022;81:36069–36127.
2. Raj S. An efficient IoT-based platform for remote real-time cardiac activity monitoring. *IEEE Transactions on Consumer Electronics* 2020;66:106–114.
3. Kwak C, Kwon OW. Cardiac disorder classification by heart sound signals using murmur likelihood and hidden Markov model state likelihood. *IET Signal Processing* 2012;6:326–334.

4. Dash TK, Mishra S, Panda G et al. Detection of COVID-19 from speech signal using bio-inspired based cepstral features. *Pattern Recognition* 2021;117:107999.
5. Al-Makhadmeh Z, Tolba A. Utilizing IoT wearable medical device for heart disease prediction using higher order Boltzmann model: A classification approach. *Measurement* 2019;147:106815–106815.
6. Guo C, Zhang J, Liu Y et al. Recursion enhanced random forest with an improved linear model (RERF-ILM) for heart disease detection on the internet of medical things platform. *IEEE Access* 2020;8:59247–59256.
7. Dash TK, Solanki SS, Panda G. Improved phase aware speech enhancement using bioinspired and ANN techniques. *Analog Integrated Circuits and Signal Processing* 2020;102(3):465–477.
8. Sengur A. An expert system based on linear discriminant analysis and adaptive neuro-fuzzy inference system to diagnosis heart valve diseases. *Expert Systems with Applications* 2020;35:214–222.
9. Ali SA, Raza B, Malik AK et al. An optimally configured and improved deep belief network (OCI-DBN) approach for heart disease prediction based on ruzzo-tompa and stacked genetic algorithm. *IEEE Access* 2020;8:65947–65958.
10. Nilashi M, Ahmadi H, Manaf AA et al. Coronary heart disease diagnosis through self-organizing map and fuzzy support vector machine with incremental updates. *International Journal of Fuzzy Systems*. 2020;22:1376–1388.
11. Dash TK, Solanki SS. Investigation on the effect of the input features in the noise level classification of noisy speech *Journal of Scientific and Industrial Research* 2019;78(12):868–872.
12. Li JP, Haq AU, Din SU et al. Heart disease identification method using machine learning classification in E-healthcare. *IEEE Access* 2020;8:107562–107582.
13. Liu Y, Guo X, Zheng Y. An automatic approach using ELM classifier for HFpEF identification based on heart sound characteristics. *Journal of Medical Systems* 2019;43:1–8.
14. Kwon JM, Kim KH, Jeon KH et al. Deep learning for predicting in hospital mortality among heart disease patients based on echocardiography. *Echocardiography* 2019;36:213–218.
15. Shuvo SB, Ali SN, Swapnil SI et al. CardioXNet: A novel lightweight deep learning framework for cardiovascular disease classification using heart sound recordings. *IEEE Access* 2021;9:36955–36967.
16. Latif S, Usman M, Rana R et al. Phonocardiographic sensing using deep learning for abnormal heartbeat detection. *IEEE Sensors Journal* 2018;18:9393–9400.
17. Wang H, Shi H, Chen X et al. An improved convolutional neural network based approach for automated heartbeat classification. *Journal of Medical Systems* 2020;44:35–35.
18. Rath A, Mishra D, Panda G et al. Heart disease detection using deep learning methods from imbalanced ECG samples. *Biomedical Signal Processing and Control* 2021;68:102820–102820.
19. Faust O, Shenfield A, Kareem M, San TR, Fujita H, Acharya UR. *Computers in Biology and Medicine*. 2018;102:327–335.

20. Wolterink JM, Leiner T, Viergever MA et al. Generative adversarial networks for noise reduction in low-dose CT. *IEEE Transactions on Medical Imaging* 2017;36:2536–2545.
21. Wang J, Liu C, Li L et al. A Stacking-based model for non-invasive detection of coro- nary heart disease. *IEEE Access* 2020;8:37124–37133.
22. Tuli S, Basumatary N, Gill SS, Kahani M, Arya RC, Wander GS, Buyya R. HealthFog: An ensemble deep learning based Smart Healthcare System for Automatic Diagnosis of Heart Diseases in integrated IoT and fog computing environments. *Future Generation Computer Systems* 2020;104:187–200.

## CHAPTER 6

# Application of Artificial Intelligence in Healthcare Sector: Benefits and Challenges

RAM SINGH[1*], ROHIT BANSAL[2], and NIRANJANAMURTHY M.[3]

[1]*School of Banking, Financial Services & Insurance Management, Symbiosis University of Applied Sciences, Indore, India*

[2]*Department of Management, Vaish Engineering College, Rohtak, Haryana, India*

[3]*Department of Computer Applications, MS Ramaiah Institute of Technology, Bangalore, Karnataka, India*

*Corresponding author

## ABSTRACT

The artificial intelligence (AI) progressions are getting ever-present in current business and ordinary everyday presence is also reliably being applied to clinical consideration, the usage of AI in clinical consideration can help clinical benefits providers in various pieces of patient thought and administrative cycles. Most AI and medical services advances have solid pertinence to the medical care field, yet the strategies they backing can differ altogether, and keeping in mind that a few articles on AI in medical care propose that the utilization of AI in medical services can perform similarly too or better than people at specific techniques, like diagnosing illness, it will be countless years before AI in medical services replaces people for an expansive scope of clinical assignments. This chapter features AI and its application in the healthcare sector, benefits,

Machine Learning Applications for Data Analysis in Healthcare Systems: Sudeshna Chakraborty, Jyotsna Singh, Praveen Kumar Shukla, Prasenjit Chatterjee (Eds.)


DOI: 10.1201/9781779643193-6

and challenges, and furthermore investigates a couple of the various sorts of man-made reasoning and medical care benefits that can be gotten from their utilization.

## 6.1 INTRODUCTION

The term "artificial intelligence" was first introduced during a gathering at Dartmouth College in 1956. Between 1956 and 1974, AI research primarily focused on solving mathematical and algebraic problems and developing systems capable of understanding natural language. From 1980 to 1987, expert systems gained popularity—these were programs designed to answer questions or solve problems based on specialized knowledge.

Interest in AI declined for a time but was revived in 1997 when IBM's Deep Blue, a chess-playing computer, defeated Russian grandmaster Garry Kasparov. Since then, nearly every major tech company and scientific institution involved in medical research or clinical trials has pursued AI-based predictive and prescriptive tools for diagnosis and treatment. According to the technology research firm Gartner, 75% of healthcare organizations were expected to invest in AI capabilities by 2021 to improve overall performance.

AI-powered medical tools offer significant benefits for both doctors and patients, and they are applicable across a wide range of healthcare domains. Beyond medicine, AI has numerous applications across industries such as finance, transportation, and manufacturing, and it is transforming how businesses make decisions and treat complex problems. AI technologies are used in areas like object, face, speech, and handwriting recognition; virtual reality and image processing; natural language processing, chatbots, and translation; e-mail spam filtering, robotics, and data mining. According to market intelligence firm Tractica, global annual AI revenue is projected to grow to $36.8 billion by 2025.

### *6.1.1 ARTIFICIAL INTELLIGENCE*

Artificial intelligence (AI) means teaching machines to think and act like humans. It can also be used to describe any machine that shows human-like features such as learning and problem-solving. AI's best quality is its ability to make decisions and take actions to achieve a goal, and a part of AI called machine learning means computers can learn and change on their own without help from people. Important ways of learning make it possible for personalized learning by taking in a lot of disorganized information like text, images, or videos.

### *6.1.2 HEALTHCARE SECTOR*

Thinking about patients in a hospital is probably the most important area, both for how much money you can make and for the amount of work you have to do. Clinical consideration includes hospitals, medical tools, basic medical care, review appointments, virtual doctor visits, medical tourism, health insurance, and medical supplies. The healthcare industry in India is growing quickly because more and more people are using it. It has two main parts: public and private. The government provides basic medical services in rural areas and some cities, while the private sector offers more advanced medical care in cities and towns. India has many skilled doctors. India is cheaper and looks different compared to other countries in Asia and the West. An operation in India is about 10 times cheaper than in Western Europe. The Indian healthcare industry is expected to reach $193.83 billion by 2025. Increasing pay, better healthcare, more people getting sick from the way they live, and more people being able to get insurance are all important factors in progress. Healthcare services are becoming stronger in India. The money made from providing medical services increased by 17.16% to $739 billion in the fiscal year 2020. According to the Economic Survey 2020–2021, the healthcare insurance industry saw more people getting insurance coverage in the fiscal year 2020. In the financial year 2020, 89% more families in Bihar, Assam, and Sikkim got medical services. In the year 2016, the country had 393 Ayurveda and 221 homeopathy government schools. In April 2020, there were 169,031 subcenters and 33,987 primary health centers. The business size was estimated to reach $372 billion by 2022.

### *6.1.3 AI AND MACHINE LEARNING*

AI is a popular type of ML used in healthcare. It is at the heart of many approaches to AI and healthcare technology, with different versions of it. Using AI in healthcare, the most common use of traditional AI is precision medicine. Being able to predict which treatment will work best for patients based on their characteristics and the treatment plan is a big step forward for many healthcare organizations. Most AI applications in healthcare, including those used in precision medicine, require large amounts of data to learn from—allowing them to accurately predict outcomes and support clinical decision-making. This is called supervised learning. AI in healthcare that uses deep learning is also used for speech recognition as natural language processing (NLP). Simplified: AI in healthcare uses deep learning for speech recognition, known as NLP. Complex deep learning models often have features that are of little importance to human observers, so it can be difficult to understand the model's results without the right interpretation.

### *6.1.3.1 NATURAL LANGUAGE PROCESSING*

Understanding human language has been a goal of computer technology and healthcare innovation for more than 50 years. Many natural language processing (NLP) systems use techniques to understand and analyze text, and then translate it. AI is used in healthcare to help understand and organize medical records. This technology can analyze patient notes to improve the quality of care and outcomes for patients (Figure 6.1).

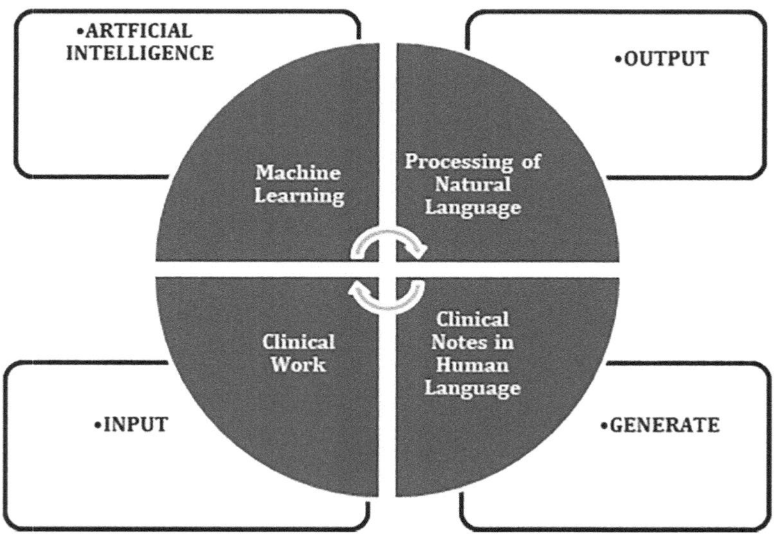


**FIGURE 6.1** NLP frameworks.

*Source*: Authors compilation.

#### *6.1.3.2 RULE-BASED EXPERT SYSTEMS*

Advanced systems rely on different sets of if-then rules as the main technology for AI in healthcare during the 1980s and beyond. AI is widely used in healthcare for helping doctors make clinical decisions even today. Many electronic health record systems (EHRs) currently provide a set of guidelines with their software offerings. Master frameworks usually require people who are experts and designers to create a set of rules in a specific area of knowledge. They work well to some extent and are easy to follow and measure. As the number of rules increases—especially beyond a few thousand—they can begin to conflict with one another, leading to system breakdowns or reduced reliability. Also, if the rules in a certain area change in a major way, it can be difficult and unfair. AI is beginning to replace rule-based systems in medical care using advanced algorithms to interpret data.

## 6.2 AI AND HEALTHCARE

Describing the work of AI in healthcare as interesting would be an understatement. AI and machine learning can make big changes in how medical care is done. There are still challenges, but the potential for this technology is so great that governments, hospitals, and providers should invest in AI-powered solutions. Starting today, AI and machine learning are ready to help healthcare providers make their work more efficient and cost-effective. Progress is making changes to healthcare services by making it easier and faster to find important information for making decisions and providing better care for patients. Using AI in healthcare can make it cheaper and better for organizations to use. According to the chief information officer, AI technology is bringing big changes to the entire patient journey in healthcare. The use of AI in healthcare focuses on making healthcare providers efficient and effective, and providing a better experience for both in-patients and out-patients. It also helps to make back-end processes more effective and organized. However, the use of AI in healthcare is not very common right now, but we hope to change that soon. Here are some ways AI and machine learning are being used in healthcare right now.

### *6.2.1 DIAGNOSIS AND TREATMENT APPLICATIONS*

For the past 50 years, AI in healthcare has focused on finding and treating illnesses. Early frameworks that followed certain standards were able to accurately analyze and treat infections, but they were not widely accepted for use in medical care. Earlier AI systems were not significantly more accurate than human clinicians in diagnosing illnesses, and they often failed to integrate smoothly with clinical workflows and electronic health record (EHR) systems. Implementing AI for diagnostic and treatment planning remains challenging due to difficulties aligning with the day-to-day practices of healthcare professionals and the technical limitations of existing EHR platforms, whether the AI tools are rule-based or algorithmic. Problems with incorporating AI into medical care have been a bigger obstacle compared to how accurate the ideas are. Most of the AI used for diagnosing and treating patients is limited to certain areas of care and comes from specific software companies. Some companies that sell EHR software are starting to add simple AI medical analysis features to their products, but it is still early. To use AI in healthcare with independent EHR systems, companies will have to either do big integration projects themselves or use outside vendors with AI skills that can work with their EHR.

### 6.2.2 ADMINISTRATIVE APPLICATIONS

AI can be used in management and patient care in hospitals. But, using AI in hospitals is not as revolutionary as using it for patient care. However, AI in hospital administration can provide important savings. Computers can support healthcare in a variety of ways, including processing insurance claims, managing patient records, and handling financial operations. AI can be used in healthcare to combine information from different databases for cases and payment management. Guarantors and suppliers need to make sure that the many cases being submitted each day are accurate. Finding and fixing coding problems and mistakes saves everyone time, money, and resources.

### 6.2.3 FACILITATING MEDICAL DIAGNOSTICS

Diagnostic errors contribute to approximately 10% of patient deaths and account for between 6% and 17% of all hospital complications. As incredibly talented as most medical care suppliers (healthcare professionals [HCPs]) are, from various perspectives, the human psyche stays uncertain.

As Andrew Beck, Director of Bioinformatics at the Beth Israel Deaconess Medical Center Cancer Research Institute, points out: "Determining the presence or absence of metastatic disease in a patient's lymph nodes is a routine yet critically important task for pathologists. Examining thousands of normal cells under a microscope to identify just a few malignant ones can be extremely tedious and challenging using traditional methods." That is the reason Beck and his group constructed a mechanized symptomatic instrument utilizing a profound learning calculation prepared to separate among destructive and noncancerous cells. In a 2016 assessment, a digital diagnostic tool achieved a success rate of 92%, only four percentage points lower than that of human pathologists. Also, when Beck's group consolidated human pathologists' investigations with the examinations of the mechanized apparatus, the indicative achievement rate increased to a noteworthy 99.5%. Ultimately, Beck believes that his analysis only begins to scratch the surface of what a hybrid approach—combining human expertise and algorithmic methods—can offer to clinical diagnostics. He states, "Our results show that what the computer is doing is quite intelligent, and that the combination of human and machine interpretations will lead to more accurate and clinically meaningful diagnoses to guide treatment decisions."

### 6.2.4 IMPROVING HEALTHCARE MARKETING

At the point when a patient has been examined, the accompanying stage is to find a treatment that will fix or potentially assuage the effects of their condition. Healthcare professionals (HCPs) clearly have a significant influence over which supportive regimen a patient adopts; however, the importance of direct-to-patient communication—such as asking about Drug X—should not be underestimated. Unfortunately, the pharmaceutical industry often ends up speaking past its core audiences. In fact, one study found that up to 45% of patients believe that drug companies do not understand their real needs. Much like the challenges faced in everyday clinical diagnostics, this disconnect is primarily a matter of scale. In the old age, monitoring patient behavior is the first step toward delivering relevant, personalized guidance. This involves collecting information from a wide range of datasets, including clinical data—such as hospital records, lab results, and healthcare provider notes—as well as general data like media preferences, online activity, and demographic information. Healthcare marketers must then extract meaningful narratives and insights from the data they have gathered—"connecting the dots," so to speak—not just once, but continuously throughout the course of a campaign. Effectively executing such a data-driven strategy at scale demands an extraordinary level of computing power. Even the most skilled and experienced marketing teams are not capable of organizing and analyzing the vast volume of data points required for effective patient targeting. This is where AI-driven predictive analytics tools become invaluable. By leveraging well-trained predictive algorithms, marketing teams can gain deep, data-driven insights into their target patient populations. These insights move beyond generalized segmentation, enabling communication strategies based on nuanced analyses of patients' past behaviors, personal characteristics, and current positions in the care journey—patterns often too complex or subtle for the human eye or mind to detect [11].

### 6.2.5 REDUCING MEDICATION NONADHERENCE

Between 1988 and 1994, approximately 38% of adults in the United States were taking at least one prescription medication. Over the following two decades, this figure rose to 49%, largely driven by a 100% increase in the number of adults taking three or more prescribed drugs. According to research presented to the American Hospital Association in October 2016, total annual spending on prescription medications had reached $309.5 billion, making it the fastest-growing component of the US healthcare economy. Alarmingly, a significant portion of this spending is wasted. It is estimated that up to 50% of the 3.2 billion prescriptions written each year in the United States are not taken as directed—if they are taken at all. Medication nonadherence results in over $250 billion in unnecessary healthcare costs annually, accounting for approximately 13% of the United States' total yearly healthcare expenditure. Just as pharmaceutical companies use algorithmic tools to enhance patient targeting, healthcare providers (HCPs) can leverage similar AI-based systems to address this issue by identifying patients most at risk of nonadherence. Factors such as socioeconomic status, insurance coverage, out-of-pocket costs, and the prescribing physician's specialty all influence a patient's likelihood of deviating from their treatment regimen. AI algorithms offer a powerful solution by analyzing these complex variables to help HCPs proactively support patients who may need additional intervention to stay on track with their prescribed therapies. Outfitted with the insightful yields of such structures, HCPs can pinpoint which patients need additional assistance to keep focused with their treatment. In reality, improved and more focused communication is not, by itself, a comprehensive solution to the problem of medication nonadherence, nonetheless, research dispersed in clinical care suggests that poor HCP-patient correspondence achieves a 19% higher threat of nonadherence.

### 6.2.6 VIRTUAL HEALTH ASSISTANT

Virtual health assistants (VHAs) can proactively help patients differently, for one, VHAs can support dementia patients by providing medication reminders, helping them stay on track with their prescribed treatments. Additionally, VHAs can offer guidance on managing common illnesses and provide personalized dietary recommendations tailored to patients' specific health needs. VHAs can also monitor patients using data-driven insights, enabling healthcare professionals to engage with both patients and pharmacies by sending reminders for medication refills and pickups. They can even recommend preventive health screenings. AI has also been employed to enhance clinical diagnostics—for instance, AI-assisted medical imaging from the Beijing-based company Infervision is currently used to improve the interpretation of CT scans and X-rays. Virtual Health Assistants represent a fusion of artificial intelligence and advanced decision-support systems, leveraging big data, NLP, and voice recognition technologies. These features contribute significantly to their ability to effectively engage and assist patients. Voice-empowered VHA will change the conveyance of medical services to patients through [9]:

- Navigate electronic health records
- Navigate diagnosis and prescription options
- Improve access to data health records
- Improve patient discharge and follow-up treatments
- Improve communication between providers

Coming up next are five different ways how VHA advances can upgrade patient commitment inside medical services associations.

#### 6.2.6.1 AUTOMATED CALLING SYSTEMS AND INTERACTIVE VOICE RESPONSE SYSTEMS (IVRS)

VHAs are increasingly being used to monitor patients before and after treatment. By receiving text and voice confirmations of upcoming appointments, VHAs help reduce cancellations and minimize outstanding bills. In clinical settings, healthcare organizations are leveraging Voice over Internet Protocol (VoIP) technologies alongside VHAs, driven by the growing adoption of AI, deep learning, NLP, and speech recognition applications on mobile devices.

#### 6.2.6.2 M-HEALTH APPS

Healthcare providers are encouraging patients to use smart devices to monitor their wellness progress, make payments, and manage many aspects of their daily routines. Some health applications available on platforms like Amazon Echo, equipped with VHAs, prompt patients to use their smartphones to request prescription refills, confirm appointments, track health metrics, and complete forms prior to visits. Wearable fitness devices, such as Fitbit, allow patients to monitor vital signs like heart rate and blood pressure. These health parameters are uploaded to patient portals, which can be easily accessed through virtual assistants.

#### 6.2.6.3 PATIENT PORTALS AND HEALTH KIOSKS

The adoption of patient health portals continues to grow at a modest pace. Ideally, these online portals—enhanced with virtual assistants—would serve as centralized platforms where patients can confirm appointments, complete medical and social history forms, update health records, view lab results, and settle outstanding balances using voice and text recognition technologies. Kiosks also hold significant potential for collecting patient information through integrated VoIP systems. Although such technologies are available in various forms, the healthcare IT infrastructure must still address key concerns related to data security and the integrity of collected information.

#### 6.2.6.4 DISCRETE DATA

VHAs can interact with patients and digitally capture their information, which can then be validated and updated in real time using cloud technologies and IoT-connected devices. When a patient engages with a virtual assistant, queries to a pharmacy hub can be cross-checked against existing records in the facility's database. Once all information is verified, it is seamlessly processed and integrated into the EHR system—eliminating the need for paper-based documentation.

#### 6.2.6.5 ANYTIME ACCESS FROM ANYWHERE

For some individuals in the maturing populace, it is undeniably more viable to have verbal correspondence alongside visual collaboration, rather than a call or electronic correspondence. VHAs can be accessed through websites, smartphones, tablets, and computers. These systems are capable of recognizing returning patients and recalling recent interactions [10].

### 6.2.7 HEALTHCARE BOTS

Healthcare bots are primarily designed to support patient engagement and accountability. These bots, often integrated into adaptive learning applications, can assist patients by delivering targeted messages and responding to health-related inquiries. They also help manage medications by providing information about drug types and recommended dosages. Recent advancements in healthcare bots include the ability to learn and mimic human conversations, detect emotions to enable empathetic interactions, and incorporate technologies such as natural language processing, sentiment analysis, and intent mining into chatbot scripts. Some bots can even perform complex image recognition tasks to analyze medical images, handwritten notes, and standardized identifiers. Beyond chatbots, other AI applications in healthcare are being developed to assess heart sounds, provide companion robots for the elderly, mine medical records, design treatment plans, assist with repetitive tasks, conduct consultations, aid in drug development, and use visual data for medical training.

## 6.3 POTENTIAL BENEFITS

AI is rapidly making its way into healthcare, playing significant roles—from automating repetitive and routine tasks in clinical practice to managing patients and medical resources. As developers design AI systems to handle these responsibilities, several risks and challenges arise. These include potential harm to patients from system errors, threats to data privacy, issues related to AI accountability, and more. Addressing these concerns requires complex solutions, including the development of better evaluation frameworks, representative datasets, community-driven oversight by regulatory bodies like the Food and Drug Administration, and reforms in medical education to prepare providers for evolving roles in a technology-driven environment. Although still an emerging field, AI is poised to play at least four major roles within the healthcare system [1].

### *6.3.1 BREAKING THE BOUNDARIES OF HUMAN CAPABILITIES*

The most significant use of clinical AI is to do things that even the most prominent healthcare providers cannot. For example, Google Health is developing a program capable of predicting severe kidney damage up to two days in advance. This early warning system contrasts with current clinical practices, which often detect such damage only after it has occurred [2]. By identifying risks earlier, this AI-driven approach has the potential to significantly enhance patient care—surpassing what is typically achievable through conventional observation alone.

### *6.3.2 DEMOCRATIZING MEDICAL KNOWLEDGE AND EXCELLENCE*

AI can also help extend the expertise and performance of specialists to providers who may otherwise lack that level of proficiency. Fields such as ophthalmology and radiology have become key areas of focus, particularly because AI-based image analysis has seen significant advancement in these domains. Some systems can analyze images of the eye to provide diagnostic insights that would traditionally require an ophthalmologist. With these tools, general practitioners—or even patients—can reach similar conclusions. This kind of democratization is crucial, especially given the shortage of highly skilled specialists in many regions [3].

### 6.3.3 THE MONOTONIC AI OF AUTOMATED MEDICAL PRACTICE

AI can automate many PC operations that now require a lot of clinical practice. Suppliers have invested a lot of energy to track electronic medical records, receive display, and perform operations on the console, even in the test room [4]. When AI systems organize key information into patient records, and facilitate documentation, narration, and coordination of data, they can significantly reduce the administrative burden on providers. This not only saves time but also enhances the provider–patient relationship by allowing more focus on meaningful clinical interactions. Ultimately, it improves the quality of care and the overall clinical experience for both parties.

### 6.3.4 PATIENT AND MEDICAL RESOURCE MANAGEMENT

Finally, AI often operates behind the scenes, yet it plays a powerful role in resource allocation and shaping healthcare operations. For example, AI systems can predict which tasks may require additional support staff, identify patients who could benefit most from limited clinical resources, or—even more controversially—optimize billing strategies to align with favorable reimbursement models.

### 6.3.5 AI FACILITATES ACCESS TO HEALTHCARE

Unfortunately, many underdeveloped countries still lag behind global technological progress in many areas, and healthcare remains their biggest challenge. People living in the most remote areas are at greater risk of being left behind, as it is difficult for them to access and manage the full range of healthcare services and technological advancements. WHO claims that lopsided admittance to wellbeing administrations brings about an 18.1-year hole in future between the most extravagant and least fortunate nations. AI innovations can play a crucial role in building an efficient and responsive healthcare ecosystem. Such digital systems help patients understand their symptoms and access appropriate treatment. In recent years, numerous applications have been developed to enhance collaboration between national and international healthcare organizations, enabling faster assistance to people in need. For instance, the

Ada app, powered by intelligent technology, is available in over 140 countries and expands access to medical guidance for vulnerable populations. Notably, Ada played a significant role during global pandemics by enabling individuals worldwide to conduct self-assessments for infectious diseases, improving early detection and response.

### 6.3.6 AI TOOLS HELP IDENTIFY DISEASE RISKS EARLY

Because AI can collect and store information about people in quarantine locations, it can use that data to view past and current illnesses. This correlation of infection complexity allows clinicians to draw more accurate conclusions. The application has arranged various indications, results, and specific cases, all of which take into account the anticipation of a person's future diseases. A notable application developed by Verily, a subsidiary of Google, specializes in diagnosing both noncontiguous and inherited genetic disorders. Such applications empower wellbeing specialists to expect possible dangers and keep away from them later on, taking proper measures today. Additionally, prescient examination improves the operational administration of medical care offices.

### 6.3.7 AI ALGORITHMS CAN SAVE TIME AND MONEY

Another great advantage of using AI in clinical examination systems is that it reduces the time and effort required to analyze and diagnose patients. Identifying biomarkers that signal the presence of certain diseases is critical to early diagnosis and treatment. Advanced computing enables the automation of much of the manual work involved in biomarker detection, significantly reducing the time required—potentially saving lives. Calculations are more useful than traditional cost estimation methods. As these technological advances are integrated into healthcare systems, patients may no longer need to undergo costly laboratory tests. As a result, there has been a significant increase—88% over the previous year—in the number of registered students completing internal audit training across various periods. This growing interest reflects the expanding role of AI in healthcare. Several AI algorithms are now proving broadly beneficial across the sector [12].

#### 6.3.7.1 NAIVE BAYES

The Naive Bayes calculation deals with the Bayes hypothesis and adopts a probabilistic strategy. In contrast to other classification algorithms, this algorithm uses a set of prior probabilities for each class. When new data is introduced, it updates these probabilities to calculate what is known as the posterior probability. This comes helpful when you need to anticipate whether the information has a place with a given rundown of classes or not.

#### 6.3.7.2 ARTIFICIAL NEURAL NETWORKS

It is a type of deep learning algorithm inspired by the neural connections in animal brains, capable of receiving signals from one layer and transmitting them to the next. These artificial neural networks can learn by analyzing patterns, often without human intervention. Their applications are diverse—ranging from aiding pathologists in diagnosis to conducting biochemical analysis. Artificial neural networks have become powerful tools across various domains in healthcare and biomedical research. Artificial neural networks are broadly categorized into two types: convolutional neural networks (CNNs) and recurrent neural networks (RNNs). Imaging plays a vital role in medical science, as it allows physicians to detect diseases even before symptoms appear. To support early diagnosis, several screening techniques—such as Pap smears, mammograms, and colonoscopies—are widely used. CNNs have proven especially valuable in this area, as they are well-suited for both binary and multiclass classification tasks in medical image analysis. In contrast, RNNs are particularly effective for pattern recognition in clinical time-series data, making them essential tools for analyzing sequences such as patient vitals or ECG readings.

#### 6.3.7.3 RANDOM FOREST

Random Forest is used to construct multiple decision trees during training for classification and regression tasks, effectively addressing the overfitting problem often associated with single decision trees. In healthcare, it is commonly applied to predict disease risk based on a patient's medical history and is also used in the analysis of ECG and MRI data.

#### *6.3.7.4 DISCRIMINANT ANALYSIS*

It is anything but an AI calculation that is utilized to dissect the ampleness of item characterization, and furthermore for appointing one item to single or numerous gatherings from the early determination of diabetic peripheral neuropathy to refining the analytic highlights of vein pictures, discriminant examination is applied in the medical services industry. It is also used in EHR management systems and for detecting signs of mental health disorders or cognitive impairments.

#### *6.3.7.5 LOGISTIC REGRESSION*

It is predominantly utilized for paired grouping. This strategy permits you to examine a bunch of factors and foresee an absolute result. Its primary applications include predicting customer lifetime value, property valuations, and other forecasting tasks.

### *6.3.8 AI AS UNIQUE AND UNRIVALLED ASSISTANCE IN SURGERY*

Effective testing and examination have been fuelling the interest in AI and advanced mechanics applications in medical procedure, and AI surgical system takes into consideration playing out the smallest and the most exact developments. Consequently, complex activities are directed with negligible agony, blood misfortune, and low dangers of results. Besides, after such surgeries, patients recuperate a lot faster. The deployment of antibacterial nanorobots helps cleanse a patient's bloodstream of infections either before or after surgery. On top of this, AI empowers surgeons with real-time data about the patient's current condition. These AI-driven tools also help build patient confidence—especially in cases where there may be anxiety or uncertainty about undergoing surgery under general anesthesia. Overall, advancements in surgical robotics are expected to significantly boost the value of the market, with projections estimating it will reach $24 billion by 2025.

### *6.3.9 AI HELPS AUGMENT HUMAN ABILITIES AND SUPPORT MENTAL HEALTH*

Patients can now receive assistance not only from clinical staff but also from robots. Technologies such as exoskeleton robots enable individuals to walk again without depending on caregivers. Another major breakthrough in AI-driven healthcare is the development of smart prosthetics equipped with sensors, allowing patients to experience limb functionality that may even surpass their original capabilities. The demand for advanced prosthetic limbs is rising, driven in part by the projected increase in the number of amputees in the United States, which is expected to reach 3.6 million by 2050, according to the Amputee Coalition.

Service robots are also playing an essential role in supporting elderly or disabled patients with daily tasks, helping reduce feelings of isolation. So-called companion and conversational robots assist with managing vital functions like temperature, blood pressure, and blood sugar levels, as well as medication reminders. Many of these robots are designed with personality analysis features to support mental health, offering emotional comfort to patients, particularly those struggling with depression.

## 6.4 CHALLENGES AND RISK

While AI offers different expected benefits, there, furthermore, are a couple of risks.

### *6.4.1 INJURIES AND ERROR*

A significant risk associated with AI in healthcare is the potential for errors that could lead to patient harm or other medical complications. For example, if an AI system recommends an inappropriate medication, fails to detect a tumor in a radiological scan, or incorrectly prioritizes patients for emergency care, the consequences can be severe. While medical errors already occur in today's healthcare systems without the involvement of AI, mistakes made by AI may raise unique concerns.

First, patients and healthcare providers may respond differently to injuries caused by software compared to those resulting from human error. Second, as AI systems become more widely adopted, a critical flaw in a single system could potentially harm a large number of patients simultaneously—unlike the more limited scope of harm caused by individual provider mistakes.

### 6.4.2 DATA AVAILABILITY

Preparing AI frameworks requires a lot of information from sources, for example, electronic wellbeing records, drug store records, protection claims records, or purchaser created data like wellness trackers or buying history; yet, wellbeing information is regularly risky and information is regularly divided across various frameworks. Indeed, even beside the assortment just referenced, patients commonly see various suppliers and switch insurance agencies, prompting information split into different frameworks and numerous configurations. This fragmentation increases the risk of errors, reduces the scope of available datasets, and raises the cost of data collection—further limiting the types of organizations that can effectively develop healthcare AI solutions.

### 6.4.3 PRIVACY CONCERNS

Another set of risks arises around patient privacy. The need for large datasets creates incentives for developers to collect extensive data from various patients, which can raise concerns about potential privacy violations. In fact, legal cases have already been filed over data-sharing practices between major healthcare systems and AI developers [5]. AI also poses unique privacy challenges: it can infer sensitive personal information that was never explicitly provided. For instance, an AI system might detect early signs of Parkinson's disease based on subtle tremors in a user's computer mouse movements—even if the individual never disclosed this information. Patients may view such inferences as breaches of privacy, particularly if those predictions become accessible to third parties such as banks or life insurance companies.

### 6.4.4 BIAS AND INEQUALITY

There are significant risks of bias and disparity in healthcare AI systems. These systems learn from the data on which they are trained, and if that data reflects existing biases, the AI can perpetuate or even amplify them. For example, if most of the training data comes from academic medical centers, the resulting AI may be less effective at treating patients from populations that rarely access such institutions. Similarly, if speech recognition AI is used to interpret clinical notes, it may perform poorly when the healthcare provider belongs to a race or gender underrepresented in the training dataset [6].

Even when AI is trained on accurate data from skilled practitioners, systemic biases can still persist. For instance, studies have shown that Black patients have historically received less treatment for pain compared to white patients [7]. An AI system trained on such records might continue to recommend lower doses of pain medication to African-American patients—not because of medical necessity, but due to ingrained systemic bias. Furthermore, AI tools used for resource allocation may exacerbate disparities by assigning fewer resources to patients deemed "less desirable" or "less cost-effective" by health systems, based on questionable criteria.

### *6.4.5 PROFESSIONAL REALIGNMENT*

AI is expected to significantly reshape the medical profession, particularly in specialties like radiology, where many tasks are automatable. While this can improve efficiency and accuracy, it raises concerns about reduced human expertise over time. As providers rely more on AI, they may lose critical skills needed to identify errors and make informed decisions, potentially weakening clinical judgment Balancing AI integration with continued human oversight is essential to maintain medical competence and ensure safe, effective care [8].

### *6.4.6 THE NIRVANA FALLACY*

One last danger bears notice. AI holds immense potential to improve healthcare, yet it faces significant risks and challenges. The nirvana fallacy arises when new technologies are unfairly compared to an idealized, flawless alternative instead of the flawed status quo. While healthcare AI is not without imperfections, inaction due to its limitations risks perpetuating the existing system's many shortcomings. Embracing AI thoughtfully may offer a path to meaningful improvement.

### *6.4.7 A LACK OF PERSONAL INVOLVEMENT*

Operational robots are thoroughly real and are not tweaked to feel any sympathy toward the patients. A couple of experts consider it to be a weight. Contrary to popular belief, human abilities—especially in personal interaction—still surpass those of computers in healthcare. The connection between doctor and patient is vital for building trust and supporting effective treatment. While AI tools can assist with routine tasks and minor errors, they lack the empathy and judgment needed in critical moments. Unlike machines, physicians can choose to go beyond strict protocols in order to save a life.

### 6.4.8 A RISE IN UNEMPLOYMENT RATES IN HEALTHCARE WORKERS

Since AI has been executed all through the whole course of action of clinical consideration on a really astonishing scale, various activities that were for the most part performed by individuals ought to be conceivable by machines nowadays. Chatbots and robots can provide emotional support, monitor patient well-being, and even predict issues like seizures, sepsis, or heart failure. While this raises concerns about job displacement, especially in routine roles, AI is unlikely to replace medical professionals in the near future. However, AI systems have already demonstrated the ability to detect diseases from X-rays with greater accuracy than some radiologists.

### 6.4.9 POSSIBILITY OF A DEFECTIVE DIAGNOSIS

Accurate diagnosis relies on data from large, diverse populations with similar symptoms. If an AI system lacks sufficient data for certain demographic groups, it may produce incorrect diagnoses. Inexperienced clinicians may not recognize these errors, leading to inappropriate treatment. Medical errors already contribute to nearly 200,000 deaths annually in the United States, with an estimated $20 billion in associated costs. Poorly trained AI tools could further increase these risks.

### 6.4.10 SOCIAL PREJUDICES

As mentioned earlier, AI-based systems cannot fully grasp human context or social nuances, which can lead to biased outcomes. For example, an AI algorithm might recommend an expensive nursing facility to a low-income patient, ignoring their financial limitations. The huge cost of organizations at such workplaces cripples the patient to continue with the treatment, and the expert can choose the best fitting treatment plan thinking about the patient's social and financial status.

## 6.5 CONCLUSION

The key challenge for AI in healthcare is not its capability, but its integration into everyday clinical practice. Over time, clinicians may shift toward tasks requiring uniquely human judgment. The true losers may be those unwilling to work alongside AI.

Despite its promise, AI cannot solve the deep-rooted issues in India's healthcare system—like inadequate funding, poor regulation, and socio-cultural barriers. While AI and the Internet of Medical Things are improving lives globally, adoption in India is hindered by weak digital infrastructure, a fragmented private sector, and uneven capacities across states and healthcare providers.

Established urban hospitals are likely to benefit first, leaving under-served areas behind. Many small clinics still rely on basic digital tools. Often, AI solutions are tech-driven rather than problem-driven, and fail to meet local needs—especially in rural regions lacking reliable internet.

Additionally, there is a growing digital divide between users and developers. Without inclusive design and robust safeguards around data security and accountability, AI in healthcare may fall short of being both safe and equitable.

## KEYWORDS

- **artificial intelligence**
- **machine learning**
- **natural language processing**
- **electronic health records**
- **chatbots**
- **healthcare**

## REFERENCES

1. Aicha AN, Englebienne G, van Schooten KS, Pijnappels M, Kröse B. Deep learning to predict falls in older adults based on daily-Life trunk accelerometry. *Sensors* 2018;18:1654.
2. Aronson S, Rehm H. Building the foundation for genomic-based precision medicine. *Nature* 2015;526:336–342.
3. Berg S. *Nudge theory explored to boost medication adherence*. American Medical Association: Chicago, 2018. www.ama-assn.org/delivering-care/patient-support-advocacy/nudge-theory-explored-boost-medication-adherence.
4. Buchanan BG, Shortliffe EH. *Rule-based expert systems: The MYCIN experiments of the Stanford heuristic programming project*. Addison Wesley: Reading, MA, 1984.
5. Char DS, Shah NH, Magnus D. Implementing machine learning in health care—addressing ethical challenges. *The New England Journal of Medicine* 2018; 378:981–983.
6. Commins J. *Nurses say distractions cut bedside time by 25 percent*. HealthLeaders, 2010. www.healthleadersmedia.com/nursing/nurses-say-distractions-cut-bedside-time-25.
7. Davenport TH, Dreyer K. AI will change radiology, but it won't replace radiologists. *Harvard Business Review* 2018. https://hbr.org/2018/03/ai-will-change-radiology-but-it-wont-replace-radiologists.
8. Davenport TH, Glaser J. Just-in-time delivery comes to knowledge management. *Harvard Business Review* 2002;;80:107-11, 126.
9. Loria K. Putting the AI in radiology. *Radiology Today* 2018;19:10. www.radiologytoday.net/archive/rt0118p10.shtml.
10. Low LL, Lee KH, Ong MEH et al. Predicting 30-Day readmissions: Performance of the LACE index compared with a regression model among general medicine patients in Singapore. *BioMed Research International* 2015;2015;169870.
11. McKinsey Global Institute. *A future that works: Automation, employment, and productivity*. McKinsey Global Institute, 2017. www.mckinsey.com/~/media/mckinsey/featured%20insights/Digital%20Disruption/Harnessing%20automation%20for%20a%20future%20that%20works/MGI-A-future-that-works-Executive-summary.ashx.
12. Vial A, Stirling D, Field M et al. The role of deep learning and radiomic feature extraction in cancer-specific predictive modelling: A review. *Translational Cancer Research* 2018;7:803–816. https://analyticsindiamag.com/top-6-ai-algorithms-in-healthcare/

# CHAPTER 7

# Design and Development of Anticancerous Drug Molecules by Structure and Ligand-Based Drug Designing Computational Approaches

PALLAVI SINGH*, SOMYA SINHA, AKASH SRIVASTAVA, and NIVEDITA UPADHYAY

*Department of Biotechnology, Graphic Era (Deemed to Be University), Dehradun, Uttarakhand, India*

*Corresponding author

## ABSTRACT

Sophistications in scientific domains have repeatedly highlighted the challenges in the drug discovery and designing pathways. Deciphering the significant lead molecules, which have potential to be effective therapeutic molecules, take time of 10–14 years and the average cost of 500–650 US million dollars when experimental techniques are deployed for drug discovery processes. In order to reduce time span, and accelerate the cost-effective approaches with high accuracy for finding novel lead molecules, computer-aided drug designing methods have been developed which make the process of hit to lead identification much easier. In this chapter, we will highlight various computational approaches, advancements in these approaches, algorithms involved in target optimization, validation and molecular docking, 2D and 3D QSAR for computer-aided drug designing. Further, we will review the inhibitors of protein–protein interactions and its design through the implementation of the pharmacophore computationally which will present its role in the discovery of the drugs. We further

Machine Learning Applications for Data Analysis in Healthcare Systems: Sudeshna Chakraborty, Jyotsna Singh, Praveen Kumar Shukla, Prasenjit Chatterjee (Eds.)


DOI: 10.1201/9781779643193-7

aim to give detailed insights on extensive databases and free online tools available to systematically conduct virtual screening, pharmacophore optimization, and drug target validation.

## 7.1 INTRODUCTION

Earlier novel drugs for known diseases were identified using random screening and observations with effectiveness of natural products. Until the 1980s, this random screening approach resulted in the discovery of numerous significant chemicals. High-throughput screening (HTS), which is ideal for independent screening of multiple drugs against a specific molecular target or cellular test, is currently improving this technique. The essence of HTS in drug discovery was emphasized with the discovery of cyclosporine A as an immunosuppressant [1]. Progressively, another successful paradigm of application of computer-aided drug design (CADD) was the finding of gefitinib using ALLADIN [2] software after screening a library of 1500 chemicals.

To unravel the mechanism of drug resistance, search for novel antibiotics, and design them for both known and unknown targets, CADD can be integrated with wet lab techniques in the design of new antibiotics. CADD approaches are computationally faster, accurate, and economical [3].

Deciphering novel targets that exhibit distinct pathways needed for bacterial life is a key option to solving antibiotic resistance. Researchers are still looking for novel antibiotics to combat established targets, and computational tools have been utilized effectively in a number of studies to do so. The fundamental CADD procedure begins with the biological identification of a hypothetical target for which ligand binding resulted in antibacterial action. In SBDD, the target's 3D structure can be determined via X-ray crystallography, nuclear magnetic resonance, or homology modeling. CADD–SBDD screening is built on this base [4]. The major focus is the development of structure–activity relationship which yields further improvements in the compounds, and LBDD is utilized in the absence of the availability of the desired 3D structure.

CADD provides for greater experiment focus, which saves time and money while discovering novel medications. The most popular and extensively used method in this area is the *in silico* methodology. These strategies are typically utilized in conjunction with the collection of *in vitro* data for constructing and evaluating the model. These models have been widely employed in the development and improvement of novel compounds with affinity for a target, as well as the elucidation of absorption, distribution, metabolism, excretion, and toxicity features, as well as physicochemical characterization.

Structure-based virtual screening (SBVS) employs scoring functions to quantify the force of noncovalent contacts between a ligand and a molecular target in order to forecast the ideal interaction mode between two molecules to create a stable complex [5]. As a result, the success or failure of SBVS software is mostly determined by scoring functions.

SBVS is performed using a variety of software packages, each of which is based on a distinct algorithm and yields various results. In the recent decade, a novel SBVS approach called consensus virtual screening has been widely employed in several researches to improve SBVS precision and minimize the number of false positives produced in these trials. The presence of a 3D structure of the target protein is an absolute need for SBVS to work. Virtual databases have been built to hold 3D structures of molecules, such as the Protein Data Bank. However, obtaining the 3D structure experimentally is not always achievable. In this case, homology modeling allows for the prediction of a protein's 3D structure based on its amino acid sequence.

## 7.2 CADD APPROACHES

1. LBDD (ligand-based drug designing)
2. SBDD (structure-based drug designing)

### *7.2.1 LBDD APPROACH*

When target structures are unknown, LBDD approaches are frequently applied in drug designing. It begins with building a ligand using software like Chemsktech, Schrodinger, and so on and generating many random combinations by adding or removing new functional groups or by Finding databases like Zinc Database to screen out a variety of ligands known to be having some biological efficacy against the target protein or gene. After building or finding the effective ligand molecule, QSAR (quantitative structure–activity relationship) approach is applied to estimate the efficacy and biological activity of those molecules. QSAR approaches involve 2D QSAR as well as 3D QSAR [6]. 2D QSAR aids in the identification and quantification of a drug's physical–chemical properties, as well as the impact of these properties on the drug's biological activity. 3D QSAR looks at a molecule's whole 3D characteristics rather than just one substituent. It revolves around a molecule's most important qualities, such as its overall size, shape, and electrical properties. Supplementation of molecules is critical in 3D QSAR. QSAR analysis is based on the fact that electronic, geometric, topological properties of the molecules govern their chemical activity, physical, and biological efficacy (i.e., its geometric, steric, and electronic properties). In QSAR, we correlate the properties of the experimentally known compounds to detect the parameters of new compounds whose efficacy we are trying to estimate. In the results of QSAR, we often used terms like biophobe and biophore. Biophore is a structure that shows some biological activity, and in contrast to it, biophobe is a chemical structure that lacks any biological activity [7]. QSAR is an amalgamation of multifarious spheres which includes statistics, chemistry, and biology. Three components must be available in order to develop a model of QSAR: (1) an experimental dataset giving insights on the properties and biological activity for a diverse group of already tested chemicals (the training set), (2) data pertaining to the molecular structure which includes many descriptors for this group of chemicals, and (3) statistical approaches which are referred to as chemometric methods for investigation and validation of the essential correlations [8] (Figure 7.1).

**IDENTIFICATION OF ACTIVE LIGANDS**

⇩

**IDENTIFICATION OD SUITABLE MOLECULAR DESCRIPTORS**

⇩

**ESTABLISHING MATHEMATICAL EXPRESSION TO CORRELATE MOLECULAR DESCRIPTORS WITH THE ACTIVITY**

⇩

**CONSTRUCTION AND VALIDATION OF THE QSAR MODEL**

**FIGURE 7.1** Flowchart explaining LBDD steps.

### 7.2.2 VIRTUAL SCREENING IN LBDD

Virtual screening is a computational technique for analyzing an extensive library of chemical compounds to find new drug candidates. It is widely used for evaluating large libraries of compounds automatically by using computer tools. Virtual screening explores the properties of the biological molecules on the basis of Lipinski's rule of five. Only those molecules which are potential to be a drug candidate based on Lipinski's rule are screened out and given in the result of virtual screening.

Virtual screening software can provide very accurate findings since they are based on quantum and molecular physics rules. We can also alter the ligands and build new forms of ligands by adding or removing specific functional groups using CHEMSKETCH software freely available online.

CHEMSKETCH is a comprehensive structure editor with a variety of tools and functionality which provide a combination of scientific and chemical information [9].

In CHEMSKETCH, we have to draw molecular structure, generate structure from InChl (The InChl program includes the ability to generate an InChl key which is a condensed 25-character representation of the InChl string. This compact representation significantly expands the use of InChl encoding for structure specification and search, particularly over the internet.) or SMILES strings, or copy/paste from chemDraw. Tools available in CHEMSKETCH allow us to incorporate the following features very easily.

- Insert predraw templates of amino acids, aromatics, steroids, sugar, and more.
- Search the dictionary of >170,000 systematic, trivial, and trade names.
- Easily draw reactions and complex chemical schemes (including biotransformation maps).
- Use graphical templates and tools to communicate chemistry and chemical biology concepts (e.g., chemical bond types Lewis's structure, molecular orbitals Newman projections, peptide sequence, and more).
- Import and export structure files in a variety of standard files and formats.

Data-based frequently explored for finding the chemical structure of known ligands and their properties are PubChem database and Zinc

Database. The PubChem database, which was launched in 2004, contains data on the biological actions of small compounds. Within the National Center for Biotechnology Information's information retrieval system, PubChem is arranged as three restricted databases. These are bioassays from PubChem [10]. PubChem also has a quick search tool for chemical structural similarities. The Zinc Database is a free virtual-screening database of commercially available chemicals. In ready-to-dock 3D formats, zinc has around 230 million comp. We can search for analogs in under a minute because zinc includes approximately 750 million purchasable compounds [11].

### *7.2.3 SBDD APPROACH*

Most of the progress has been done in the field of molecular biology and structural advancement has been made for the determination of the spectroscopic structures. These methodologies have obtained the three-dimensional structures of more than 1 lakh proteins [12]. Besides the storage and retrieval of data, there has been a sufficient development of sophisticated techniques for the analysis of biological data computationally. Improvement in bioinformatics and Human Genome Project completion has paved out the way for the development of drugs due to the accessibility of innumerable target proteins. Three-dimensional structures of those proteins that are known to be therapeutic in nature are favoring the binding cavities identification and have formed the basis for drug designing according to its structure. SBDD is a speedy process for the lead optimization and discovery as it governs the three-dimensional structure of the receptor protein and houses the information about the illness at a molecular basis [13]. Most of the computational techniques like molecular simulations, molecular docking, and SBDD are most commonly used in SBDD.

SBDD is a recursive process as it progresses through more than one cycle leading to an enhanced drug molecule to medical tribunals. Generally, any detection prior to the drug involves four steps: phase 1: discovery; phase 2: development; phase 3: clinical; phase 4: registry. During the first phase, a potent healing target and a dynamic ligand is being recognized. Numerous computational systems can be utilized for binding the vast records of the data comprising small molecules into the binding site of the receptor protein. Ranking of the molecules is performed rendering the system of scoring on the basis of the electrostatic forces with the binding site. In-depth examination of the properties of the electrostatic forces of attraction for the binding site, with the existence of cavities, allosteric pockets, and clefts can be accomplished by utilizing the three-dimensional structure of the molecule to be targeted. Currently used SBDD methods contemplate the significant characteristics of the binding cavity of the therapeutic target for the competent designing of novel ligands [14, 15]. The second phase involves the synthesis and optimization of top hits [16]. Moreover, the compounds which are ranked on the top with high kinship for discerning intonation of the protein which is to be targeted are examined *in vitro* in the biochemical assays. These molecules as ligands interfere with the vital cellular pathways, thus leading to the development of drugs with the anticipated pharmacological and therapeutic effects [17]. Experimental methods are used for the evaluation of the biological properties like affinity, potency, and efficacy of the discerning complexes [12]. The subsequent step involves the determination of the three-dimensional structure of the receptor protein in complex with the capable ligand attained in the first phase. The three-dimensional structure delivers comprehensive evidence about the intermolecular forces responsible for the molecular recognition and ligand binding. Structural acumens into the ligand and protein complex aids with the analysis of several binding conformities, binding pockets identification of the unidentified ones, and ligand–receptor protein interaction; explication of the alterations in the conformation which results from the ligand binding; and thorough mechanistic studies [18]. In due course, multiple recurrences intensify the usefulness and meticulousness of the potential drug molecule. The third phase includes the clinical trials of the developmental candidates which are further taken into consideration. The compounds which pass the clinical trial phase are then further being taken to the last phase for the distribution of the drug in the market subjected for the clinical use.

### *7.2.4 STRUCTURE PREDICTION TECHNIQUES EMPLOYED IN SBDD*

Many techniques are used in structure prediction during SBDD, and it includes comparative modeling (also known as homology modeling), threading, and ab initio modeling. The technique used on a wider scale for predictive modeling of three-dimensional structure of proteins is homology modeling [19]. In homology modeling, we predict the structure of the unknown query sequence based on selecting a template protein whose structure is known and has 40% or above sequence similarity.

Homology modeling is very useful in predicting protein structure when only sequence data is available. When a target protein of an unknown structure has a structural homolog in the database, the structure of the target protein can be modeled by using a template. This method is composed of six basic steps:

- Template selection: FASTA and BLAST are used for sequence alignment to find the template protein for the target protein. Minimum 35% sequence similarity should be there.
- Target-template sequence alignment: After finding a template sequence, alignment of the query sequence with the template sequence is carried out by using software T-coffee and PRALINE [20].
- Model building: It has two substeps known as the backbone model building and loop modeling.
- After model building, Side-chain refinement is done using algorithms like SCWRL.
- Built model is optimized by model refinement procedures. Model refinement is done by using energy minimization functions. We use software like GROMOS for model refinement.
- Model evaluation is carried out by PROCHECK and WHAT IF.

After the prediction of three-dimensional structure of the target, it is essential to correlate the model by checking the stereochemical properties and this is done by assessing the prediction structures using Ramachandran plot [21]. This helps in determining torsion angles $\psi$ and $\varphi$ and help in providing insight about the peptides structure. After applying Ramachandran plot, predicting the active site of the target protein where the ligand will bind, is an essential step. Multiple freely available tools and software are used for active site or binding cavity residues prediction. These methods are based on the evaluation of the interaction energy and van der Waals forces. Q-site finder is one such tool which is used for active-site prediction. Q-site finder is an energy-based method commonly used for binding-site prediction [22]. In this method, van der Waals interaction energies of proteins are calculated with a methyl probe. Those that have favorable energy are employed.

These probe clusters are placed accordingly to their total interaction energies additionally interacting protein residues are functionally annotated to determine the binding site [23].

### 7.2.5 ONLINE TOOLS FOR IMPLEMENTATION OF SBDD

The compound database preparation containing small molecules is frequently accessible via synthesis or purchase, which owns necessary characteristics like solubility, stability in aqueous media, presence of apposite functional groups for the interaction with the target bioactive compounds, and the absenteeism of noxious and unwanted moieties. Various rules have been applied for ensuring the "likeliness of the drug" with the "Lipinski's rule of five" [24] which states that the drug compound should have a molecular weight less than 500, log *P* value which is the lipophilicity to be lesser than 5, must possess less than five hydrogen bond donor and less than 10 hydrogen bond acceptors. Further, ADMET studies which is an amalgamation of selecting the chosen pharmacological distribution, absorption, excretion, metabolism, and toxicologic properties which is clearly advised in the earlier stages of the drug design process. In order to efficiently screen the collection of compounds against the said criteria, various online servers have been technologically evolved.

Chembioserver is one of the publicly available online applications which aids in the filtering and assortment of molecules which are smaller in size [25]. The aim of this online application is in the preparation of the compound by utilizing the main segments of it including basic search, filtration on the basis of toxicity, advance filtering, clustering, modified pipeline, and use of the property graphs for the visualization of the properties of the compound which will thus upsurge the efficacy and the compound's quality that proceeds to *in vitro* assays. The FAF-Drugs2 server is a tool which is publicly available for filtering the compounds computationally by keeping in view the physiochemical properties and identifying the vital functional and unwanted moieties [26]. This tool delivers various diagrams on distribution related to the properties of the selected compounds in web-based server version [27].

The datasets of the compounds should be first processed in the three-dimensional representations in the realistic view. Moreover, the compound which is set to be used for SBVS should have angle and bond lengths as they may not be reformed during the process of docking. Moreover, it must be bereft of escorting wreckages like solvent molecules, metals, and counter ions, all other compounds are required to have a valency, the state of protonation at an appropriate pH, and the state of tautomerism [28,29]. The programs associated with docking include the tools of Autodock software for the preparation of ligands [30], MAPS [31], molecular operating environment [32], LigPrep, HyperChem, and Pipeline plot [34]. A workflow has been created by Grinter et al. [34] for preparing a repository for performing the operation of docking which includes scripts for the preparation of ligands which are made freely available to the academic communities [35].

Last and most crucial step in CADD in both approaches is molecular docking. We will understand molecular docking, the parameters of molecular docking, algorithms involved in energy minimization, and online tools available to conduct molecular docking in the next section of this chapter.

## 7.3 MOLECULAR DOCKING

Molecular docking refers to fitting in the potential ligands in the active site of the target protein and generates the least energy, best-fit molecular conformation. Various molecular parameters are analyzed during docking followed by multiple iterative rounds to minimize the energy using algorithm like Newton–Raphson, steepest descent, and conjugate variate methodologies [36]. Molecular docking generates plethora of docked structures and conformations which are then categorized using scoring function feature. Scoring function is based on multiple parameters such as free energy, topological properties, electronic structure, stability of complexes, binding energy, root mean square deviation, and so on [37].

The accurate ligand-binding conformation depends on two rules which are as follows:

- Conformation space which helps in finding suitable binding regions.
- Binding energy of each prediction is correlated with each conformation.

Molecular docking is a multiphase process and is done when a state with minimum energy is achieved, in which ligand binding is checked by various scoring functions. The goal of molecular docking is to examine feasible binding of genome having specific ligand with a target whose

3D structure is known [38]. There are three basic features that need to be accomplished in docking:

- depiction of the binding site,
- ligand fixing with the specific binding region, and
- assessing the energy of interaction between a specific ligand–receptor complex [39].

Pharmacophore modeling is required as a preliminary step in many cases before going for molecular docking. Pharmacophore is a 3D arrangement of functional groups having a molecular structure. Pharmacophore modeling refers to generating a model of the specific conformation of active functional groups which is derived from a set of known ligands that are selected for a target [40]. This model depends upon steric and electronic features. It uses algorithms that encompass the following steps: input, conformational search, feature extraction, representation of structure, pattern identification, and scoring [41]. It is generally used to select hits for docking studies. Scoring functions are mathematical tools by which binding affinity of two molecules is predicted after they are docked.

### 7.3.1 DOCKING PARAMETERS

Docking has a large number of input parameters based on its functionality. We need to check accessible surface area, force field parameters acting on each molecule, dipole moment, stearic complementarity between receptor and ligand, physicochemical properties of the ligand and receptor. Docking can be carried by varying different parameters listed below:

- Flexible ligand: Used to relax an input conformation; have parameters such as anchor search, anchor size, and torsion drive [42].
- Orient ligand: Helps in orientation search; done by matching.
- Score ligand: Does scoring of orientation.
- Minimize ligand: It minimizes conformations and improves the efficiency.
- Multiple ligand
- Chemical screen: Allows rapid filtering of data based on 3D structure.
- Random seed: Figure to seed the random number generator [43].

### *7.3.2 DOCKING ALGORITHMS*

Docking uses different search algorithms, which help in predicting possible confirmation of complexes. These algorithms are based on certain parameters. On the basis of flexibility of the receptor and ligand, the search algorithms are divided into two types.

*Rigid body docking*: In this algorithm, molecular conformation remains rigid and ligand is fitted in by varying the orientation of ligand in various angles. It is also the first slant used in docking and it compile the below-mentioned functions such as it fixes the protein and ligand and helps in selecting molecules with moderate energy orientation and protein–protein docking is done by this.

*Flexible docking*: It is the most used form of docking. This ligand is flexible but the receptor is rigid. First ligands are selected to become flexible by moving firm parts through optimization. In this bond angle, torsion angles are modified and internal geometry is changed.

- Systematic search algorithms provide variations in the structure, and rapidly change ligand confirmation. Examples of some systemic search algorithms are conformational search, database, fast shape mappings, and so on. It is further categorized into fragment and exhaustive search algorithms. Exhaustive searches clarify ligand forms by systematically rotating all possible bonds and torsion angles in a given period of time. The fragment-based algorithm is done by incremental construction of ligand conformations from fragments obtained by selecting ligands of interest [44].
- The Monte Carlo algorithm creates a casual initial configuration of ligand in the active site which is scored on the basis of energy and other parameters. After this, small new changes are made and the novel confirmation is again subjected to scoring. The better confirmation with higher scoring is selected, also known as the initiation minimization program [45].
- Genetic algorithms are built on the principle of population and biological evolution [46]. There is a set of parameters that decides the arrangement of protein and ligand based on translation, rotation, and confirmation.
- Tabu search algorithm in this small change is made to the current configuration of ligand and on the basis of fitness, they are ranked [47].

### 7.3.3 TOOLS AND SOFTWARE APPLIED FOR MOLECULAR DOCKING

Different programs use different docking algorithms. Most of the programs are based on virtual screening. The following are major programs used in docking:

1. AutoDock: It is a modeling simulation software which is very effective in protein–ligand docking. It is an automated tool that helps in predicting how small fragments adhere to receptors of known 3D structure. It also has applications in chemical mechanism studies, X-ray crystallography, virtual screening, and SBDD. It consists of two programs, one is to DOCK and another is AUTOGRAID [48,49]. URL: http://autodock.scripps.edu/
2. AutoDockVina: This is a free molecular docking program and is a new generation of Autodock with improvement in speed and accuracy level. It internally calculates the grid for the atom types and precalculating grid maps for them. It has applications in lead optimization, protein docking, and so on. URL: http://vina.scripps.edu/
3. DOCK: It is a program for multiple protein structures and uses a similar matching algorithm to build ligand binding orientation [50, 51]. It uses sphere ligand-matching algorithm of UCSF DOCK to generate confirmations.
4. Glide: It is a ligand-docking program and it predicts protein–ligand binding. It is best and feasible in doing HTS [52]. URL: http://www.schrodinger.com/glide/
5. FlexX: It is a high-throughput software [10,53–55]. URL: http://www.biosolveit.de/FlexX/

## 7.4 APPLICATIONS OF CADD IN THE DEVELOPMENT OF ANTICANCEROUS DRUGS

Computational methods act as a boon in the modern era for discovering new drugs. As it has been known to cover all the hierarchy of the drug discovery, it is found to be applicable in the curing of one of the serious diseases which is cancer as it requires less labor, less resources, and is usually a less time-taking process. More often, computational methodologies have become a promising tool in most of the cases related to the development of drugs for the treatment of cancer. Here, the listing of various fruitful applications of the computational approaches for small molecules, which have been found to be functional for the treatment of cancer or are being utilized at the later on stages for clinical trials.

Crizotinib is one of the major examples of the drug developed by using the concept of the SBDD [56, 57]. This drug has been regarded as a discerning and powerful c-Met dual inhibitor, which was Food and Drug Association (FDA) approved in the year 2011 [58]. Hepatocyte growth factor, a natural ligand, and its receptor have been known to play an imperative role in cell signaling [59]. There has been a significant amount of upregulation of the protein called as c-Met which has been perceived in the carcinoma of several patients [60] and an improper working of the c-Met signaling was detected in the solid and blood tumorigenic cancer. Thus, it has been proved that c-Met is a potent and striking target of oncology.

One of the FDA-approved drugs, Axitinib, has acted as a novel therapy for the treatment of renal cell carcinoma (RCC) [61]. The development of this drug was possible by utilizing the strategy of SBDD in which Axitinib served as an inhibitor on binding to the vascular endothelial growth factor (VEGF) kinase domain in the DFG phenylalanine configuration [62]. The family of VEGF has its occupancy in the signaling process which is further involved in the angiogenesis process. The signaling pertaining to VEGF was detected in the lump of the cells, and this process of signaling finds its role in the development of diseases that are in the malignant form. VEGF receptor serves as the ligand in the VEGF signaling process. These receptors are called as the family of tyrosine kinases called as $FLT_1$, KDR, $FLK_1$, and $FLT_4$. Blocking the mechanism of action of these receptors with the pan-AKT kinase inhibitor against the kinases family have been proven to be an effective manner of developing drugs against angiogenesis. More recently, Inlyta®, Axitinib in association with KEYTRUDA®; pembrolizumab, has been approved as the drug against RCC [63]. Moreover, NVP-AUY922; Luminespib has been known to possess a highly strong activity for the treatment of tumors in the carcinoma of lungs thereby directing mutations like insertion and models at the cellular level in an assenting trial.

## 7.5 ADVANTAGES OF *IN SILICO* DRUG DESIGNING OVER EXPERIMENTAL TECHNIQUES

CADD has far more superior outcomes as compared to conventional wet lab techniques. CADD techniques provide a higher ratio of identifying compounds with the desired properties and increasing chance of a compound overcoming the barriers of preclinical testing. It reduces the need for animal testing and has shortened the timespan required for HTS of millions of molecules. Structure- and ligand-based methods are the predominantly applied tools in the drug discovery domain but the future lies in the upcoming synergistic techniques such as proteochemometrics. Combinatorial chemistry and proteomics have been infused to lead to remarkable results in the CADD era. Integration of omics technologies and high-scale accelerated development of machine learning algorithms has reduced the timespan of processing. Sophistications in cloud computing, data mining, data management, big data, and artificial intelligence have strengthened the data storage and data retrieval capacities in biological databases leading to accentuated virtual screening and molecular simulations. Metagenomics and metatranscriptomics have made it possible to analyze bulk environmental samples in one go and further add up to the availability of bulk-level genomic sequences and transcriptomes available on the servers. CADD is way ahead in drug designing and future lies in the biological data processing and integration with high-end computational technologies.

## 7.6 CONCLUSION

The future of drug discovery and development lies in higher advancements in CADD techniques related to virtual screening, lead identification, optimization, molecular docking, pharmacophore modeling, and molecular modeling. Identification of target structure, validation, and screening of lead molecules has become easier with available databases and tools like 2D and 3D QSAR. Life-threatening diseases like cancer can be combated in a better way using these advancements. There is still a need to further develop simulation algorithms with better efficacy than molecular dynamics simulation and Monte Carlo approaches.

## KEYWORDS

- **high-throughput screening**
- **computer-aided drug design**
- **structure-based virtual screening**
- **ligand-based drug designing**
- **quantitative structure–activity relationship**
- **molecular docking**

## REFERENCES

1. Kania R. Structure-based design and characterization of axitinib. In Li R, Stafford JA (Eds.), *Kinase Inhibitor Drugs, Drug Discovery and Development*. Wiley: New York, 2009.
2. Kapetanovic IM. Computer-aided drug discovery and development (CADDD): In silico-chemical-biological approach. *Chemical Biology Interact* 2008;171:165–176. https://doi.org/10.1016/j.cbi.2006.12.006
3. Blaney J. A very short history of structure-based design: How did we get here and where do we need to go? *Journal of Computer-Aided Molecular Design* 2012;26:13–14. https://doi.org/10.1007/s10822-011-9518-x
4. Kitchen DB, Decornez H, Furr JR, Bajorath J. Docking and scoring in virtual screening for drug discovery: Methods and applications. *Nature Reviews Drug Discovery* 2004;3:935–949. https://doi.org/10.1038/nrd1549
5. Sousa SF, Fernandes PA, Ramos MJ. Protein -ligand docking: Current status and future challenges. *Proteins* 2006;65:15–26. https://doi.org/10.1002/prot.21082
6. Alonso H, Bliznyuk AA, Gready JE. Combining docking and molecular dynamic simulations in drug design. *Medical Research* 2006;26:531–568. https://doi.org/10.1002/med.20067
7. De Vivo M, Masetti M, Bottegoni G, Cavalli A. Role of molecular dynamics and related methods in drug discovery. *Journal of Medicinal Chemistry* 2016;59:4035–4061. https://doi.org/10.1021/acs.jmedchem.5b01684
8. Lima AN, Philot EA, Trossini GH, Scott LP, Maltarollo VG, Honorio KM. Use of machine learning approaches for novel drug discovery. *Expert Opinion on Drug Discovery* 2016;11:225–239.
9. Wilson GL, Lill MA. Integrating structure-based and ligand-based approaches for computational drug design. *Future Medicinal Chemistry* 2011;3:735–750. https://doi.org/10.4155/fmc.11.18.
10. Ferreira LG, dos Santos RN, Oliva G, Andricopulo AD. Molecular docking and structure-based drug design strategies. *Molecules* 2015;20:13384–13421. https://doi.org/10.3390/molecules200713384
11. Anderson AC. The process of structure-based drug design. *Chemical Biology* 2003;10:787–797. https://doi.org/10.1016/j.chembiol.2003.09.002.
12. Ferreira LG, Dos Santos RN, Oliva G, Andricopulo AD. Molecular docking and structure-based drug design strategies. *Molecules* 2015;20(7):13384–421. https://doi.org/10.3390/molecules200713384. PMID: 26205061; PMCID: PMC6332083.
13. Lionta E, Spyrou G, Vassilatis DK, Cournia Z. Structure-based virtual screening for drug discovery: Principles, applications and recent advances. *Current Topics in Medicinal Chemistry* 2014;14(16):1923–38. https://doi.org/10.2174/1568026614666140929124445. PMID: 25262799; PMCID: PMC4443793.
14. Mandal M, Boese B, Barrick JE, Winkler WC, Breaker RR. Riboswitches control fundamental biochemical pathways in *Bacillus subtilis* and other bacteria *Cell* 2003;113:577–586.
15. Wilson GL, Lill MA. Integrating structure-based and ligand-based approaches for computational drug design. *Future Medicinal Chemistry* 2011;3(6):735–750.

16. Urwyler S. Allosteric modulation of family C G-protein-coupled receptors: From molecular insights to therapeutic perspectives. *Pharmacological Reviews* 2011;63(1):59–126.
17. Zhao Y, Wang L, Jin Y, Shi S. Fas ligand regulates the immunomodulatory properties of dental pulp stem cells. *Journal of Dental Research* 2012;91(10):948–954.
18. Lipinski CA. Avoiding investment in doomed drugs. *Current Drug Discovery Technologies* 2001;1:17–19.
19. Lionta E, Spyrou G, Vassilatis DK, Cournia Z. Structure-based virtual screening for drug discovery: Principles, applications and recent advances. *Current Topics in Medicinal Chemistry* 2014;14:1923–1938.
20. Kalyaanamoorthy S, Chen YP. Structure-based drug design to augment hit discovery. *Drug Discovery Today* 2011;16:831–839. https://doi.org/10.1016/j.drudis.2011.07.006.
21. Tollenaere JP. The role of structure-based ligand design and molecular modelling in drug discovery. *Pharmacy World & Science* 1996;18(2):56–62.
22. Plewczynski D, Lazniewski M, Augustyniak R, Ginalski K. Can we trust docking results? Evaluation of seven commonly used programs on PDB bind database. *Journal of Computational Chemistry* 2010;32(4)742–755, https://doi.org/10.1002/jcc.21643.
23. Jones G, Willett P, Glen RC, Leach AR, Taylor R. Development and validation of a genetic algorithm for flexible docking. *Journal of Molecular Biology* 1997;267(3):727–748
24. Athanasiadis E, Cournia Z, Spyrou G. ChemBioServer: A web-based pipeline for filtering, clustering and visualization of chemical compounds used in drug discovery. *Bioinformatics* 2012;28(22):3002–3003. https://doi.org/10.1093/bioinformatics/bts551. Epub 2012 Sep 8. PMID: 22962344.
25. Lagorce D, Sperandio O, Galons H, Miteva MA, Villoutreix BO. FAF-Drugs2: Free ADME/tox filtering tool to assist drug discovery and chemical biology projects. *BMC Bioinformatics* 2008;9:396. https://doi.org/10.1186/1471-2105-9-396. PMID: 18816385; PMCID: PMC2561050.
26. Lagorce D, Maupetit J, Baell J, Sperandio O, Tufféry P, Miteva MA, Galons H, Villoutreix BO. The FAF-Drugs2 server: A multistep engine to prepare electronic chemical compound collections. *Bioinformatics* 2011;27(14):2018–20. https://doi.org/10.1093/bioinformatics/btr333. Epub 2011 Jun 2. PMID: 21636592.
27. Sadowski J, Rudolph C, Gasteiger J. The generation of 3D-models of host-guest. *Analytica Chimica Acta* 1992;265:233–241. https://doi.org/10.1016/0003-2670(92)85029-6.
28. Kalliokoski T, Salo HS, Lahtela-Kakkonen M, Poso A. The effect of ligand-based tautomer and protomer prediction on structure-based virtual screening. *Journal of Chemical Information and Modeling* 2009;49(12):2742–8. https://doi.org/10.1021/ci900364w. PMID: 19928753.
29. Morris GM, Huey R, Lindstrom W, Sanner MF, Belew RK, Goodsell DS, Olson AJ. AutoDock4 and AutoDockTools4: Automated docking with selective receptor flexibility. *Journal of Computational Chemistry* 2009;30(16):2785–91. https://doi.org/10.1002/jcc.21256. PMID: 19399780; PMCID: PMC2760638.
30. MAPS, version 3.4. Paris, France: Scienomics SARL; 2014.

31. Canada, H3A 2R7: Montreal, QC. Molecular Operating Environment (MOE) St. West, Suite #910. 2013.
32. LigPrep. New York, NY: Schrödinger, LLC; 2013.
33. Hyperchem. HyperCube, Gainesville, FL.:
34. Grinter SZ, Yan C, Huang SY, Jiang L, Zou X. Automated large-scale file preparation, docking, and scoring: Evaluation of ITScore and STScore using the 2012 community structure-activity resource benchmark. *Journal of Chemical Information and Modeling* 2013;53(8):1905–14. https://doi.org/10.1021/ci400045v. Epub 2013 May 21. PMID: 23656179; PMCID: PMC3755023.
35. Kung PP, Jones RA, Richardson P. *Crizotinib (Xalkori): The First-in-Class ALK/ROS Inhibitor for Non-small Cell Lung Cancer*. John Wiley & Sons, 2015.
36. Jiang F, Kim SH. "Soft docking": Matching of molecular surface cubes. *Journal of Molecular Biology* 1991;219(1):79–102.
37. Alonso H, Bliznyuk AA, Gready JE. Combining docking and molecular dynamic simulations in drug design. *Medicinal Research Reviews* 2006;26(5):531–568.
38. Subramanian J, Sharma S, B-Rao C. Modeling and selection of flexible proteins for structure-based drug design: Backbone and side chain movements in p38 MAPK. *ChemMedChem* 2008;3(2):336–344
39. Diller DJ, Merz KM Jr. High throughput docking for library design and library prioritization. *Proteins* 2001;43(2):113–124
40. AskMayoExpert. *Cervical cancer (adult)*. Mayo Foundation for Medical Education and Research: Rochester, Minn., 2019.
41. Lobo RA et al. Malignant diseases of the cervix. In *Comprehensive Gynecology.* 7th edn. Philadelphia, PA: Elsevier; 2017. https://www.clinicalkey.com. (accessed June 14, 2019).
42. Niederhuber JE et al. eds. Cancers of the cervix, vulva and vagina. In *Abeloff's Clinical Oncology*. 6th edn. Elsevier: Philadelphia, PA, 2020. https://www.clinicalkey.com. (accessed June 14, 2019).
43. AskMayoExpert. *Tamoxifen and CYP2D6 (pharmacogenomics)*. Mayo Foundation for Medical Education and Research: Rochester, MN, 2017.
44. Toledo E et al. Mediterranean diet and invasive breast cancer risk among women at high cardiovascular risk in the PREDIMED trial. *JAMA Internal Medicine* 2015;175:1752.
45. Breast cancer prevention—Patient version (PDQ). National Cancer Institute. https://www.cancer.gov/types/breast/hp/breast-prevention-pdq. (accessed June 28, 2017).
46. Pruthi S et al. Successful implementation of a telemedicine-based counseling program for high-risk patients with breast cancer. *Mayo Clinic Proceedings* 2013;88:68.
47. Pruthi S (expert opinion). Mayo Clinic: Rochester, MN, 2017.
48. Warner KJ. *Allscripts EPSi*. Mayo Clinic: Rochester, MN, 2019.
49. Anders CK et al. Epidemiology, risk factors and the clinical approach to ER/PR negative, HER2-negative (triple negative) breast cancer. https://www.uptodate.com/contents/search. (accessed May 3, 2019).
50. Flint PW et al. Malignant neoplasms of the oral cavity. In *Cummings Otolaryngology: Head & Neck Surgery.* 6th edn. Saunders Elsevier: Philadelphia, PA. 2015. https://www.clinicalkey.com. (accessed Nov. 28, 2018).

51. Born J, Manica M, Oskooei A, Rodriguez Martínez M. *PaccMannRL: Designing Anticancer Drugs from Transcriptomic Data via Reinforcement Learning.* Cornell University Press: New York, 2019
52. Chan HCS, Wang J, Palczewski K, Filipek S, Vogel H, Liu ZJ et al. Exploring a new ligand binding site of G protein coupled receptors. *Chemical Science* 2018;11. https://doi.org/10.1039/C8SC01680A
53. Chan HCS, Shan H, Dahoun T, Vogel H, Yuan S. Advancing drug discovery via artificial intelligence. *Trends in Pharmacological Sciences* 2019;592–604. https://doi.org/10.1016/j.tips.2019.06.004
54. Chen X, Yan CC, Zhang X, Zhang X, Dai F, Yin J, et al. Drug-target interaction prediction: Databases, web servers and computational models. *Briefings in Bioinformatics* 2016;696–712. https://doi.org/10.1093/bib/bbv066
55. Eskiler GG. Talazoparib to treat BRCA-positive breast cancer. *Drugs Today.* 2019;459–467. https://doi.org/10.1358/dot.2019.55.7.3015642
56. Cui JJ, Tran-Dube M, Shen H, Nambu M, Kung PP, Pairish M, et al. Structure based drug design of crizotinib (PF-02341066), a potent and selective dual inhibitor of mesenchymal-epithelial transition factor (c-MET) kinase and Anaplastic Lymphoma Kinase (ALK). *Journal of Medicinal Chemistry* 2011;6342–6363. https://doi.org/10.1021/jm2007613
57. Cui JJ, Mctigue M, Kania R, Edwards M. Case History: XalkoriTM (Crizotinib), a potent and selective dual inhibitor of Mesenchymal Epithelial Transition (MET) and Anaplastic Lymphoma Kinase (ALK) for cancer treatment. *Annual Reports in Medicinal Chemistry* 2013;421–434. https://doi.org/10.1016/b978-0-12-417150-3.00025-9
58. Christensen JG, Burrows J, Salgia R. C-Met as a target for human cancer and characterization of inhibitors for therapeutic intervention. *Cancer Letters* 2005;225:1–26. https://doi.org/10.1016/j.canlet.2004.09.044
59. Liu X, Yao W, Newton RC, Scherle PA. Targeting the c-MET signaling pathway for cancer therapy. *Expert Opinion Investigational Drugs* 2008;17:997–1011. https://doi.org/10.1517/13543784.17.7.997
60. Meadows KL, Hurwitz HI. Anti-VEGF therapies in the clinic. *Cold Spring Harbor Perspectives in Medicine* 2012;2:27. https://doi.org/10.1101/cshperspect. a006577
61. Kania RS, Bender SL, Borchardt AJ, Cripps SJ, Hua Y, Johnson MD, et al. *Indazole compounds and pharmaceutical compositions for Inhibiting protein kinases, and methods for their use.* 2016
62. Atkins MB, Plimack ER, Puzanov I, Fishman MN, Mcdermott DF, Cho DC, et al. Axitinib in combination with pembrolizumab in patients with advanced renal cell cancer: A non-randomised, open-label, dose-finding, and dose-expansion phase 1b trial. *Lancet Oncology* 2018;19:405–415. https://doi.org/10.1016/S1470-2045(18)30081-0
63. Rong B, Yang S. Molecular mechanism and targeted therapy of Hsp90 involved in lung cancer: New discoveries and developments (Review). *International Journal of Oncology* 2018;52:321–336. https://doi.org/10.3892/ijo.2017.4214

# CHAPTER 8

# Smart Healthcare Systems: An Exigency of Current Era

ANUPAMA SHARMA[1], PRASHANT SRIVASTAVA[1], PRATEEK CHAURASIA[1], SHWETA ROY[2*], and SANDHYA AVASTHI[2]

*[1]Department of Information Technology, Ajay Kumar Garg Engineering College, Ghaziabad, Uttar Pradesh, India*

*[2]Department of Computer Science and Engineering, ABES Engineering College, Ghaziabad, Uttar Pradesh, India*

*Corresponding author

## ABSTRACT

Smart healthcare systems are the exigency of the present healthcare situation. Health services can be administered in a systematic and successful manner with smart healthcare records. Healthcare systems integrating smart blockchain, artificial intelligence enhancement, and Internet of Things (IoT) support help doctors to devote more time to patient care and reduce their need to search for clinical information. In today's world, patients need to travel abroad to see their doctors for better treatment but during treatment they face many issues; smart healthcare records may help to reduce such issues. Within the healthcare framework, blockchain network serves as the platform for storing and exchanging patient data across hospitals, diagnostic labs, pharmaceutical firms, and medical practitioners. Blockchain applications enable the accurate identification of critical errors fraudulent activities. It presents an exclusive data storage framework that ensures maximum security, alleviating concerns regarding data tampering. This chapter presents a review on the exigency

Machine Learning Applications for Data Analysis in Healthcare Systems: Sudeshna Chakraborty, Jyotsna Singh, Praveen Kumar Shukla, Prasenjit Chatterjee (Eds.)


DOI: 10.1201/9781779643193-8

of smart healthcare systems in the current era, the technologies involved to build such smart healthcare systems, current research work to produce smart healthcare systems, and further research directions to enhance such systems for further benefits. Researchers may benefit from analyzing further research directions in this field.

## 8.1 INTRODUCTION

Human health conditions are crucial to maintain and necessary to improve human pleasure and social progress. It is also essential to the advancement of the economy of the country. Every day, accidents, illnesses, and crises take place and the treatment and diagnosis of the disorders are anticipated. A health record is a compilation of clinical information regarding the patient's physical and mental health, as determined by various sources. Nowadays, a patient's medical history, current medications, examination, diagnosis, treatment, and lab test reports, warnings about allergies, and so on need to be maintained both manually and digitally. Smart healthcare systems will be able to transform it totally in digit prints.

### *8.1.1 EXIGENCY OF SMART HEALTHCARE SYSTEMS*

The provision of healthcare is vital to life. There must be a method to ease the burden on the healthcare system. This industry has been slow to adopt new technologies because of the complexity of the system. Due to the enormous influence that even the tiniest acts may have in this industry, as well as the sheer number of stakeholders and invested parties, every move must be carefully considered. However, as the world continues to improve with smart technology, healthcare has also turned into a proactive adopter of technology. Due to the significant contribution of advanced technology, increasing patient safety and lowering healthcare costs such systems are now becoming the current need. The healthcare sector may benefit from quickly expanding research and technology since it is primarily data driven. In some cases, doctors all around the world use pens to make handwritten notes and store the information offline in massive stacks, which make it difficult to access important information when needed. Such situations created requirements for smart/advanced healthcare systems.

### 8.1.2 ISSUES AND CHALLENGES ASSOCIATED TO SMART HEALTHCARE SYSTEMS

Design and implementation of a smart healthcare system may face various issues and challenges. Some most challenging points are as follows:

1. *Funding*: The smart healthcare system has a number of challenges, including funding. IoT infrastructures are expensive and difficult to acquire, which makes their implementation challenging. A smart healthcare system also costs money to maintain.
2. *Bureaucracy*: Bureaucracy stands as a significant challenge in the smart healthcare system. Bureaucracy in healthcare refers to the complex administrative structures, regulations, and processes that can sometimes impede the efficiency of healthcare systems. While bureaucracy serves the purpose of ensuring compliance, quality control, and standardization, an excessive bureaucratic framework in healthcare can lead to various challenges. Changes in organizational and cultural norms are frequently more challenging than changes in technology.
3. *Healthcare practitioners' limited computer proficiency*: The absence or lack of computer proficiency among healthcare practitioners can affect the adoption of telemedicine, data analytics, and other digital healthcare initiatives. Without adequate computer knowledge, healthcare professionals may struggle to efficiently interpret and use digital health data or leverage technology-driven patient management tools.
4. *Intermittent power supply*: The success of remote healthcare applications may be hampered by an epileptic power supply. Inconsistent power supply poses a significant challenge to successful remote healthcare delivery. The irregularity in power availability could disrupt crucial healthcare services, affecting patient consultations, medical device functionality, and data transmission. This unpredictability impedes efforts to provide consistent and reliable healthcare services to remote or underserved areas.
5. *Security and privacy*: This is a severe problem in IoT-based healthcare systems because numerous devices and protocols are used to access critical patient information that is stored in the cloud. Security and privacy in healthcare are paramount, ensuring the protection of sensitive patient data from unauthorized access, breaches, or misuse.

### *8.1.3 PIVOTAL INFRASTRUCTURE*

It is possible to monitor the physiological conditions of the body with wearable sensors. Vital indications including body temperature, heart rate, and breathing rate can all be measured by the sensors. It is possible to add more blood pressure and blood oxygen sensors. The information from the sensor nodes is processed by a central node before being sent outside to another site. A specialized central node that only performs one function is preferred to reduce power consumption.

Healthcare is only now beginning to use artificial intelligence (AI) and related technologies. AI has already been proven to function at least as well as humans at critical healthcare tasks. AI is made up of a variety of distinct technologies. Digital data can be shared and recorded via blockchain technology without being altered.

## 8.2 CURRENT STATUS OF RESEARCH AND DEVELOPMENT IN SMART HEALTHCARE SYTEMS

Lots of research work is going on in the field of smart healthcare systems and their development. Chen et al. [1] worked on handling the emergency situation of patients and proposed the edge-cognitive-computing-based (ECC-based) smart healthcare system. Using cognitive computing, this system can track and examine the physical health of its users. Additionally, it adjusts each user's health-risk grade when allocating computer resources throughout the entire edge computing network. The results of the trials demonstrate that the ECC-based healthcare system offers a better user experience, efficiently utilizes computing resources, and dramatically increases patient survival rates in an unexpected emergency. Uddin [2] suggested a smart system to monitor the patient's activities without hospitalization and it is possible with the use of new technologies such as the Internet of Things (IoT), Edge of Things, and Cloud of Things. Recurrent neural network (RNN) technology on an edge device is used in this work to present a wearable sensor-based system for activity prediction (i.e., personal computer or laptop). The system's input data come from a variety of wearable healthcare sensors, including magnetometer, accelerometer, gyroscope, and electrocardiography sensors. An RNN is then trained using the features. The RNN that has been trained is utilized to forecast the activities. In order to provide older persons, patients, and individuals with disabilities with a simple control system, Ismail et al. [3] presented an effective method based on voice recognition. To quickly access

IoT devices put in smart homes and hospitals without relying on a centralized supervisory system, a low-cost speech recognition system is being developed. The suggested system employed a Raspberry Pi device to provide wireless smartphone management of home appliances. This system's major objective is to make it easier for users to connect with household appliances using IoT communications that are based on voice commands. The proposed framework improves voice recognition by combining a hybrid support vector machine with a dynamic time-warping algorithm. Ahad et al. [4] presented a survey on the use of IoT in smart healthcare systems. Tripathi et al. [5] examined the social and technological impediments to the adoption of smart hospital system (SHS) by examining user perception and cutting-edge expert opinions. Haque et al. [6] suggest a smart healthcare system to resolve the problems of current healthcare systems. Smart contracts built on blockchain technology are used to protect confidential data and uphold privacy. Smart contracts enable a considerable number of actions to be performed automatically. As a result, the system becomes more resource-friendly because it depends less on physical labor. It has been established that the information kept is accurate, unchangeable, and always necessitates proper authority approval. Additionally, healthcare has become universally accessible, safe, decentralized, and automated. The use of machine learning techniques for big data analysis in the healthcare industry has been thoroughly reviewed by Li et al. [7] in their research. Additionally, the strengths and weaknesses of the currently used approaches as well as numerous research difficulties are emphasized. Their research will help government organizations and healthcare professionals stay up to date on the latest developments in machine learning–based big data analytics for smart healthcare. Li et al. [8] proposed a federated learning-based special three-layer (i.e., user, client, cloud, etc.) architecture, ADDetector which achieves privacy-preserving by addressing privacy breaches that exist in data, feature, and model levels. Sujith et al. [9] developed blockchain, deep-learning-based healthcare system. The blockchain framework's integration improved patient confidentiality and data privacy, preventing data misuse against patients. Syeda et al. [10] discussed smart healthcare and its applications, they covered the significance and function of developing technologies, including AI, the IoT, blockchain, 3D printing, and 5G networking (AIB35). Hameed et al. [11] proposed a theoretical autonomous healthcare model that separates hospital–patient collaborations from the overall healthcare system. The suggested concept makes use of numerous blockchains, including a global blockchain consortium (GBC), which is used for hospital collaboration building a global healthcare system where a patient is not constrained to a certain group of hospitals and physicians located in a particular region. For small issues that do not need much doctor engagement or consultation, other blockchains (regional blockchain consortium [RBC]) are tailored to the regions of the patients. RBCs are permissioned, but the GBC is a permissionless blockchain. The suggested paradigm promises a decentralized framework that makes it easier for users and other parties to share, exchange, and integrate information.

## 8.3 TECHNOLOGIES AND ARCHITECTURE TO IMPLEMENT SMART HEALTHCARE SYSTEMS

Machine learning and IoT open whole new doors for healthcare. The collection and analysis of physiological data from anybody, anywhere, at any time has been made possible by the development of transformational technologies. The implementation of smart healthcare systems involves various technologies and architectural frameworks. These systems encompass cutting-edge advancements such as artificial intelligence (AI), IoT, blockchain technology, and cloud computing. The combination of these technologies creates an intricate architecture to revolutionize healthcare delivery.

AI-driven algorithms analyze patient data, enabling accurate diagnoses and personalized treatment plans. IoT devices, including wearable sensors, monitor vital signs, fostering continuous health tracking. Blockchain technology ensures secure and immutable data sharing among healthcare stakeholders, maintaining patient privacy and data integrity. Cloud computing facilitates vast data storage and seamless access to information across healthcare networks.

The synergy of these technologies forms a robust architecture enhancing healthcare services. It enables remote patient monitoring, streamlines administrative tasks, and facilitates predictive analytics for better health outcomes. Embracing these advancements in healthcare architecture promises improved efficiency, accessibility, and quality of medical services for patients.

### *8.3.1 INTERNET OF THINGS*

Wearable sensors are available to monitor bodily physiological states. Technology allows a central node to process the data from the sensor nodes and transfer it to another location outside. To minimize power usage, a specialized central node that only performs one task is preferable.

All sensors are possibly compact, transportable, and wearable through advanced technology. Patients now have a nonintrusive, comfortable solution that can track their health continuously. Patients may be more open to employ health monitoring technology through smart healthcare systems. Such systems can be made more useful if cameras or implantable sensors are attached to take updated information about patients. Short-range connections, like Bluetooth, are recommended by current models for sending sensor data to smartphones for processing. The patient's processed information can subsequently be transferred to the healthcare practitioner, usually a doctor, using SMS or the Internet using long-range connections like long-term evolution. The main drawback of using cell-phones for healthcare updates is the battery life. Cellphones frequently need to be recharged because of their short battery life; a patient with a dead battery would not be connected to health practitioners.

Patient physiological data are monitored and collected using sensors and the patient is identified using a radio frequency identification (RFID) tag. This IoT-based RFID healthcare monitoring device allows for the collection of physiological data from hospital patients and elderly people. Ambient sensors can detect the physical parameters of the surroundings, like the presence of toxic agents, humidity, and temperature could be used to assess the environmental parameters and analyze their relation with the person's health (Figure 8.1).

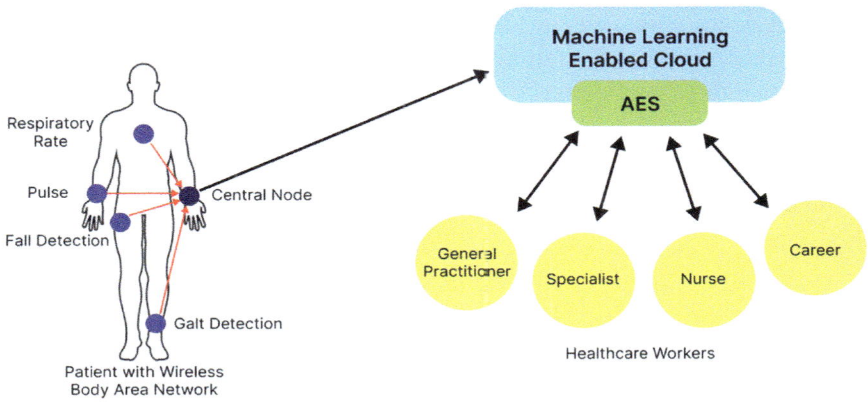


**FIGURE 8.1** Smart healthcare system framework.

IoT can be used for the formation of a SHS, which relies specifically on RFID, wireless sensor network, and smart mobile, interoperating with each other over low-power wireless personal area network. The SHS is able to gather environmental conditions and patients' physiological parameters in real time through an ultra low-power hybrid sensing network made up of network nodes incorporating ultrahigh-frequency RFID functions. Sensing data are sent to a control center, where a sophisticated monitoring program makes it available via a REST web application programming interface (API) for both local and remote users.

Medical information about patients must be adequately protected to prevent unauthorized individuals from accessing sensitive data about patients. The majority of earlier studies on IoT applications in healthcare delivery did not examine how crucial it is to protect patient's medical information stored in the cloud from security flaws. Advanced encryption standard (AES) is a symmetric block cipher that is used to encrypt sensitive data and is deployed in hardware and software all around the world to secure classified information.

### *8.3.2 ARTIFICIAL INTELLIGENCE AND MACHINE LEARNING*

The application of AI and associated technologies in healthcare is just taking a good shape. AI has already been shown to perform at least as good as humans for crucial activities in healthcare. There are several different technologies that makeup AI in healthcare systems.

Low-power nodes like mobile phones of patients can be used to transmit data collected from the sensors on the patient's body over to the cloud. Machine learning algorithms can be deployed in the cloud itself using this enormous data, which will be produced quickly and will continue to increase in cloud storage. These algorithms might be created to shift through vast volumes of data, spot patterns in diseases that were not recognized before, and offer diagnoses and treatment options, among other things.

### *8.3.3 BLOCKCHAIN TECHNOLOGY*

Blockchain makes it possible to share and store digital data while preventing its change. The limitations of the current healthcare systems are fixed by the smart healthcare system. Blockchain-based smart contracts are used to preserve privacy and safeguard sensitive data. Numerous tasks can be carried out automatically via smart contracts. As a result, the system uses fewer resources because it is less reliant on manual labor. It has been proven that the data are reliable, unalterable, and always require the proper authority's approval. Additionally, healthcare has become automated, decentralized, safe, and available to everyone.

The GBC aims to lead all blockchain-pertinent activities on a single, well-equipped platform. GBC can be used for healthcare collaborations also. GBC is able to establish a worldwide healthcare system that frees patients from geographic restrictions on their choice of hospitals and doctors. Other blockchains known as RBC are adapted to the locations of the patients for minor ailments that do not require a lot of doctor interaction or advice. RBCs are permissioned, but the GBC is a permissionless blockchain. The suggested paradigm promises a decentralized framework that makes it easier for users and other parties to share, exchange, and integrate information (Figure 8.2).

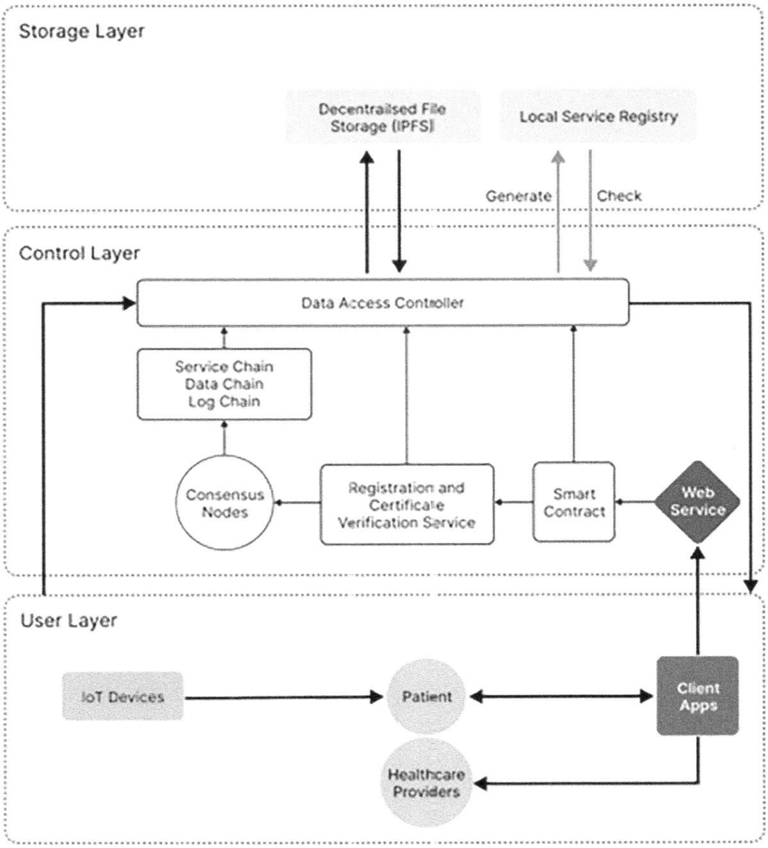


**FIGURE 8.2** Blockchain architecture in healthcare.

## 8.4 HEALTHCARE USE CASES

The healthcare field is drastically enhancing with technological advancement and according to the needs of current situations of the patients. Patients require smart care of their self or for their relatives who want special care of their patients during their busy schedules. Advanced technologies such as IoT, blockchain, AI, big data, and so on are promising technology to take over the current situation of healthcare systems. Smart healthcare systems have been developed and available as applications for mobile and another electronic gadget to keep a track of diagnosis and updates about health conditions. Quick and easy follow-ups are possible with doctors and better quality treatment facilities can be availed by patients through such smart healthcare systems. Authentic patient records will be available for further treatment procedures and patients will be able to avail hassle-free health services due to such advanced technology. Some patients may be disabled or have physical limitations and require wearable devices to assist their body or to get updates regarding body parameters. Smart healthcare systems have various healthcare use cases, which are as follows.

### *8.4.1 HEALTHCARE SUPPLY CHAIN MANAGEMENT*

Healthcare supply chain management ensures that hospitals, clinics, and pharmacies have the right things at the right time. It involves lots of steps, like ordering, storing, and delivering medical stuff, while making sure everything is safe and of good quality. Various kinds of integration that define health supply chains are as follows:

- Process coordination and integration.
- Coordinating and integrating information flow.
- The blending and coordinating of planning procedures.
- The blending of internal and external organizational processes.
- Integration of a market-based strategy.
- Market development integration.

#### *8.4.1.1 TRACEABILITY OF DRUGS*

In the administration of the healthcare supply chain, blockchain transactions are a crucial monitoring tool for the entire process of moving medicines and medical products. Due to the fact that every transaction is recorded on the ledger and updated by every node in the blockchain, it is easy to instantly validate the drug's origin, the vendor, and the distributor. Additionally, the distributed ledger of the blockchain enables officials and medical professionals to confirm and validate the credentials of providers. In this sense, the development of a reliable vendor network that enables healthcare managers to protect patients from shady vendors shows significant promise for blockchain technology. Furthermore, blockchain technology promises to significantly improve transactional security, data provenance, demand forecasting, and fraud prevention. Using blockchain technology, Figure 8.3 depicts a pharmaceutical supply chain management procedure.

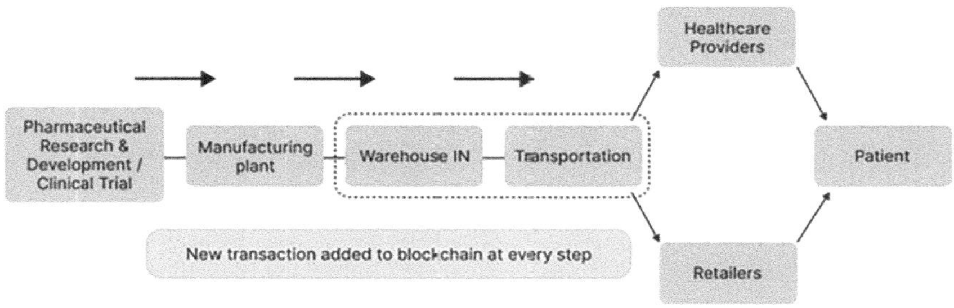


**FIGURE 8.3** Pharmaceutical supply chain management procedure.

Step 1: When a new treatment or medication is developed, a process begins involving patent protection and extensive clinical testing. These data are recorded in a digital ledger, serving as a transaction entry.

Step 2: After completing clinical trials, the manufacturing facility obtains a patent for prototype testing and large-scale production. Each product receives a unique identification linked to additional blockchain data.

Step 3: Post mass production and packaging, medicines are stored in warehouses for future distribution. The blockchain records details like timestamps, lot numbers, barcodes, and expiry dates.

Step 4: Transportation specifics, such as travel duration between warehouses (*in*), mode of transport, authorized agents, and related particulars, are also logged in the blockchain.

Step 5: Third-party distribution networks transport drugs and medical supplies to healthcare providers or retailers. Utilizing *in* and *out* warehouses linked to all distribution endpoints, blockchain updates with new transactions.

Step 6: Healthcare facilities like hospitals and clinics input data (batch and lot numbers, product owners, and expiry dates) into the blockchain to authenticate and prevent counterfeiting.

Step 7: Stores follow similar procedures as Step 6 for authentication and validation within the blockchain.

Step 8: With the blockchain providing transparent information for verification, patients are empowered to ensure authenticity throughout the entire supply chain process.

#### *8.4.1.2 SECURITY OF DRUGS*

The supply chain participants would be encouraged to conduct better testing, maintain better temperature control, and ensure the authenticity of their drugs by a blockchain's transparency. This is because the blockchain tracking technology would make it possible for direct chain players. As a result, it would be easy to identify the location in the chain, for example, when one chain member committed fraud.

According to Schindler, "If I have any product, I can attach a virtual token to it, which is stored on a blockchain." Therefore, this token must be given with the product. Let us imagine a business manufactures 100 units of a product, each of which has a digital token kept in a blockchain. For instance, they are given to the wholesaler, and this transaction is recorded in the blockchain. The wholesaler then transfers it to a different distributor, and once more, this transaction is recorded. If the distributor, for example, adds 200 copies of the product falsely, the blockchain would make it clear that only 100 copies were made, and the fake ones would not have a token attached to them. Therefore, it would be quite simple for any recipient of those shipments to doubt the legitimacy of the bogus goods. Additionally, the originating company might reconsider its business relationships in order to eliminate the fraudulent distributor from its supply chain.

In addition to maintaining the integrity of the supply chain, this visibility and ability to trace products might also enable pharmacists or even patients to view the data directly, giving them confidence in their medications.

### *8.4.2 MAINTENANCE OF RECORDS TO FACILITATE THE HEALTHCARE STAFF*

Efficiently maintaining records significantly aids healthcare staff in delivering quality care. Streamlining record-keeping processes through electronic systems allows swift access to patients' medical histories, aiding in accurate diagnoses and tailored treatments. Digital records enable healthcare professionals to monitor a patient's progress, medications, and allergies seamlessly. Moreover, centralized databases ensure information availability across healthcare facilities, enhancing coordination among medical teams. Efficient record maintenance optimizes administrative tasks, reducing paperwork, and enabling staff to focus more on patient care. Embracing advanced technologies like electronic health records (EHR) or management information systems amplifies productivity, ensuring prompt and precise documentation. This organized data repository facilitates informed decision-making and efficient communication among healthcare providers, elevating the overall quality of healthcare services.

### 8.4.3 PLATFORM TO EXCHANGE PATIENT'S RECORDS

A platform for sharing patient records is like a digital space where hospitals, doctors, and clinics can swap important information about a person's health. It works a lot like a big, safe, and secret locker for your health data that only authorized people can access. This locker keeps all your medical details safe and organized, like your test results, past illnesses, and prescribed medicines. When you go to different doctors or hospitals, they can open this locker to see what has happened before. It is like having a special book about your health that all your healthcare helpers can read to understand your story. This sharing helps everyone involved in your treatment to work together better. They can understand your health history, plan treatments, and ensure you get the right care. It is like a sharing library for your health information that only trusted people can use to help you feel better.

### 8.4.4 AUTOMATED SYSTEMS TO ASSIST DISABLES

Blockchain makes it possible for health records to be used properly by reducing needless administrative expenses. Additionally, the necessity for several middlemen to oversee the sharing of crucial health information will decline as a result of this technology. Delivering efficient, timely, and sufficient services to their people is the main duty of healthcare professionals. Thanks to blockchain technology, healthcare providers will already be aware of each patient's medical background. Where blockchain provides security and transparency, IoT and artificial intelligence provide auto control to smart healthcare systems which may be really able to assist disabled persons. Wearable computing is the main attraction of this use case of healthcare.

## 8.5 SMART HEALTHCARE SYSTEMS DESIGN ASPECTS

Smart healthcare systems are crucial to design. Such systems have a variety of use cases and have many difficulties in the collaboration of involved advanced technologies. The following design aspects are necessary to take care while designing of smart healthcare systems.

### *8.5.1 COLLECTION AND MANAGEMENT OF DATA*

Data gathered from the sensors on the patients' bodies can be sent to the cloud using low-power nodes, such as their mobile phones. Machine learning algorithms can be deployed in the cloud itself using this enormous data, which will be produced quickly and will continue to increase in cloud storage. This gathered data will quickly take the form of big data and various big data tools would be required for handling it.

### *8.5.2 SECURITY FEATURES*

Stringent access controls, including multifactor authentication, restrict system entry to authorized personnel only, enhancing data confidentiality. AI-powered threat detection continuously monitors system activities, promptly identifying and mitigating potential security threats or irregularities.

Adherence to stringent regulatory frameworks such as Health Insurance Portability and Accountability Act and General Data Protection Regulation ensures compliance with data protection laws, prioritizing patient privacy and rights. Privacy-preserving techniques embedded within AI algorithms anonymize patient data during analysis, preserving confidentiality.

Secure APIs facilitate safe data exchange between different healthcare platforms or systems. Regular security audits and assessments identify vulnerabilities, ensuring system resilience against evolving cyber threats. Continuous training programs for healthcare staff on cybersecurity practices mitigate human error risks and unauthorized access.

### *8.5.3 PRIVACY ISSUES*

Strictly using symmetric keys, medical IoT data and EHR are kept private. Only the encrypted healthcare data hash is present throughout the entire chain. Adversaries can only obtain encrypted data from nodes as a result.

Healthcare data are secured with a symmetric key that is additionally encrypted with the authorized user's public key from a 2048-bit Rivest–Shamir–Adleman key pair. The healthcare data cannot be obtained by adversaries without the symmetric key. As a result, this plan offers a high level of data privacy.

### *8.5.4 INFORMATION EXCHANGE*

The paradigm in Figure 8.4 promises a decentralized framework that facilitates information sharing, exchange, and integration between users and other parties. Patient's health information sharing with other concern expert requires lots of carefulness. Smart healthcare systems provide security and privacy to such sensitive patient's health-related reports and other analyses.

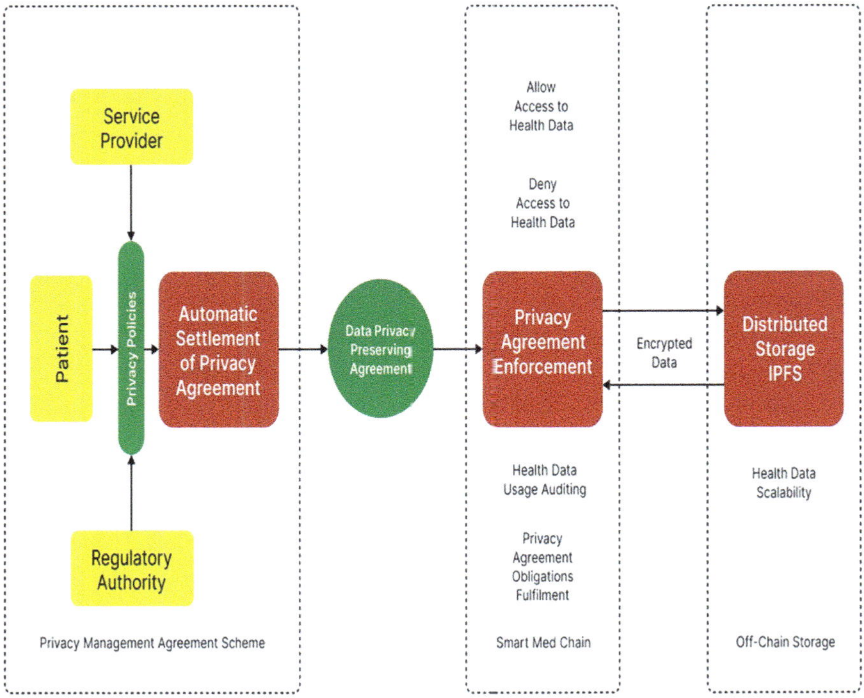


**FIGURE 8.4** Implementation of data privacy and security in smart healthcare.

## 8.6 FUTURE TRENDS FOR SMART HEALTHCARE SYSTEMS

Future of smart healthcare systems has the following trends:

### *8.6.1 INCREASING USE OF TELEHEALTH SERVICES*

Telehealth offers a spectrum of services, including virtual consultations, remote patient monitoring, and telemedicine applications, fostering timely and convenient access to medical expertise. Its utilization spans various healthcare domains, ranging from primary care to specialty consultations, mental health services, and chronic disease management. Telehealth, facilitated by digital technologies, enables remote access to healthcare services, consultations, and monitoring, transcending geographical barriers. Particularly catalyzed by the COVID-19 pandemic, its popularity continues to soar due to its convenience, accessibility, and efficiency.

### *8.6.2 DIGITAL HEALTH TO THE RESCUE*

It uses technology to make things easier for everyone. Imagine having a doctor's appointment online using your phone or computer instead of going to the clinic. Digital health also includes apps that help keep track of your fitness, remind you to take medicines, and even connect you with a doctor instantly in case of an emergency. It is like having a smart helper that keeps an eye on your health all the time. Moreover, it allows doctors to share information quickly, ensuring you get the best care possible. With digital health, people living in faraway places can also consult doctors without traveling long distances. Wearables and smart gadgets are able to monitor a patient's every action, including their heart rate, calorie intake, sleep pattern, and activity duration. Wearable technology, for instance, can spot abnormal electrocardiographic patterns in patients and notify medical staff, lowering the risk of more catastrophic illnesses. They have enabled the methodical advancement of personalized healthcare, and they will continue to be crucial to the advancement of medical devices in 2022 and furthermore years.

### *8.6.3 IMPROVEMENT IN DATA SCIENCE AND PREDICTIVE ANALYTICS*

Data science advancements empower healthcare by deciphering scanned images to detect bodily anomalies, aiding doctors in devising effective treatment strategies. Handling copious data from patients with chronic ailments poses challenges, yet it grants doctors profound insights. This includes identifying high-risk patients for potential illnesses and gathering comprehensive patient information. The evolution of data science and predictive analytics facilitates intricate tasks, enabling efficient analysis of vast medical datasets. By leveraging these technologies, healthcare professionals gain deeper insights into patient conditions, optimize treatment plans, and predict potential health risks.

### *8.6.4 IMPACT OF ARTIFICIAL INTELLIGENCE*

AI is reshaping healthcare profoundly. It revolutionizes diagnostics by enhancing accuracy in interpreting medical images, aiding in timely and precise diagnoses. Operations in healthcare facilities are streamlined through AI-driven automation, freeing up healthcare providers to focus more on patient care. AI's ability to analyze extensive patient data enables personalized treatments and predictive analytics, foreseeing potential health risks. Moreover, it accelerates drug discovery and development, speeding up the introduction of new medications and therapies. Telemedicine solutions powered by AI facilitate remote patient monitoring, ensuring access to quality care, especially in remote areas. Its role in healthcare research is pivotal, analyzing vast datasets to identify patterns, and fueling medical breakthroughs.

### *8.6.5 THE CYBERSECURITY MESH*

The Cybersecurity Mesh is an approach that aims to provide more flexible, adaptable, and responsive security measures across various digital environments. Rather than relying on a centralized security perimeter, the Cybersecurity Mesh focuses on creating a more dynamic and distributed security infrastructure. It does this by decentralizing security measures and extending protection to individual devices, people, applications, and services, regardless of their location. In the past, it was shown that more than 34% of the data were vulnerable to leaks and breaches within the network itself when security was focused all around the perimeter. With the Cybersecurity Mesh, many businesses anticipate improved cybersecurity.

This security architecture was expected to be able to lower the cost effect of a security incident by 90%, according to some study.

### 8.6.6 CLOUD COMPUTING

Cloud computing in healthcare revolutionizes data storage and management, offering cost-effective on-demand services. It reduces capital expenditure, enhances operational efficiency, and simplifies tasks like sharing medical records and developing telehealth applications. The cloud ensures secure, scalable, and accessible storage, facilitating seamless collaboration among healthcare providers while streamlining processes for improved patient care and optimized resource utilization.

### 8.6.7 BLOCKCHAIN

Every time a linked device participates in a transaction of any kind, the blockchain system will produce identical blocks. A block is created on all devices to store the data locally whenever data on one computer is accessed, updated, shared, or otherwise processed. The medical industry has suffered greatly as a result of the inability to access and exchange sensitive patient data securely. But for true interoperability, blockchain will provide incredibly flexible openness while preserving the strictest security standards.

### 8.6.8 VIRTUAL, AUGMENTED, AND MIXED REALITY

Healthcare has seen significant technological advancements, especially with the adoption of virtual reality (VR) and augmented reality. These technologies hold immense potential for the healthcare sector in India. VR has shown promise in aiding vision rehabilitation, addressing conditions like cancer, autism, and depression. In mixed reality, the merging of virtual and physical realms offers clinicians a unique opportunity to analyze patient conditions more comprehensively. These advanced educational tools enable healthcare professionals to provide more precise treatments.

## 8.7 CONCLUSION

Healthcare technology is continuously becoming advanced in every aspect. Despite the industry's improvements in security, evolving threats persist, hence necessitating a focus on prevention rather than reaction. This proactive approach is crucial to effectively manage these challenges. The quality and effectiveness of healthcare are expected to continue its improvement because of support and the evolution of the latest technologies like AI, machine and deep learning, and so on. Smart diagnosis facilitated by these technologies allows for a more accurate definition of a patient's condition and disease status. This, in turn, enables the creation of personalized treatment plans. With the expert endorsement, the software can guide precise treatment pathways. It is imperative to collaborate with skilled software developers who understand the system's requirements and objectives to achieve these advancements.

## KEYWORDS

- **Internet of Things**
- **blockchain network**
- **smart healthcare systems**
- **healthcare professionals**
- **artificial intelligence**

## REFERENCES

1. Chen M, Li W, Hao Y, Qian Y, Humar I. Edge cognitive computing based smart healthcare system. *Future Generation Computer Systems*. 2018;86:403–411.
2. Uddin MZ. A wearable sensor-based activity prediction system to facilitate edge computing in smart healthcare system. *Journal of Parallel and Distributed Computing*. 2019;123:46–53.
3. Ismail A, Abdlerazek S, El-Henawy IM. Development of smart healthcare system based on speech recognition using support vector machine and dynamic time warping. *Sustainability*. 2020;12(6):2403.

4. Ahad A, Tahir M, Aman Sheikh M, Ahmed KI, Mughees A, Numani A. Technologies trend towards 5G network for smart health-care using IoT: A review. *Sensors*. 2020;20(14):4047.
5. Tripathi G, Ahad MA, Paiva S. S2HS—A blockchain based approach for smart healthcare system. In *Healthcare*. Elsevier. 2020, Vol. 8 (1), p. 100391.
6. Haque AB, Muniat A, Ullah PR, Mushsharat S. An automated approach towards smart healthcare with blockchain and smart contracts. In *2021 International Conference on Computing, Communication, and Intelligent Systems (ICCCIS)*, IEEE. 2021, pp. 250–255.
7. Li W, Chai Y, Khan F, Jan SRU, Verma S, Menon VG, Li X. A comprehensive survey on machine learning-based big data analytics for IoT-enabled smart healthcare system. *Mobile Networks and Applications*. 2021;26(1):234–252.
8. Li J, Meng Y, Ma L, Du S, Zhu H, Pei Q, Shen X. A federated learning based privacy-preserving smart healthcare system. *IEEE Transactions on Industrial Informatics*. 2021;18(3).
9. Sujith AVLN, Sajja GS, Mahalakshmi V, Nuhmani S, Prasanalakshmi B. Systematic review of smart health monitoring using deep learning and artificial intelligence. *Neuroscience Informatics*. 2022;2(3):100028.
10. Syeda MZ, Syeda DEH, Babbar H. The role of emerging technologies in smart healthcare. *IoT-Enabled Smart Healthcare Systems, Services and Applications*. 2022;1–17.
11. Hameed G, Singh Y, Haq S, Rana B. Blockchain-based model for secure IoT communication in smart healthcare. In: *Emerging Technologies for Computing, Communication and Smart Cities*. Springer: Singapore, 2022, pp. 715–730.

CHAPTER 9

# Technology-Enabled Smart Healthcare toward Smart Society 5.0

CHABI GUPTA[1], FINCY PALLISSERY[2], and KRITI PARASHAR[2]

[1]*Amity College of Commerce and Finance Amity University, Noida, India*

[2]*School of Law, Christ University Delhi NCR Campus, Delhi, India*

*Corresponding author

## ABSTRACT

The emerging technologies such as artificial intelligence (AI), the Internet of Things (IoT), blockchain, 3D printing, and 5G networking (referred as AIB35) in the context of "smart healthcare" are being greatly explored by "smart healthcare" providers to move forward in an all-encompassing manner. The field of healthcare requires scientists to implement intelligent solutions into the daily routines of "smart aging" patients to improve the efficiency and precision of diagnostic and therapeutic procedures. The concept of "smart aging" refers to a program that makes it possible for elders to have a superior quality of life by embracing high-end technology solutions to diagnosis and disease. Healthcare technologies are a wide range of innovations, including connected devices, appliances, and procedures, as well as drugs, medicine, and vaccine that improve the level and quality of care provided at economic cost structures. The purpose of this study is to first present a listing of the principal technologies that underpin "smart healthcare," and then to present it in several significant domains. After that, we elaborate on the issues that now exist with smart healthcare and attempt to provide potential solutions to these issues. In the final part of this chapter, we take a glance into the road ahead and assess the prospects of smart healthcare toward building a smart society 5.0.

Machine Learning Applications for Data Analysis in Healthcare Systems: Sudeshna Chakraborty, Jyotsna Singh, Praveen Kumar Shukla, Prasenjit Chatterjee (Eds.)


DOI: 10.1201/9781779643193-9

## 9.1 INTRODUCTION AND LITERATURE REVIEW

The age that we live in is the age of information. Traditional medicine has gradually begun to get digital especially in recent years because of research-oriented advances in technology and scientific theory. There has been an emergence of intelligent and smart healthcare that makes use of the most recent inventions of information technology [9]. Smart healthcare is not simply a simple technology improvement; rather, it is also a transformation that affects healthcare on a comprehensive and multilevel scale [17]. Changes in the healthcare model (moving away from diagnosis to prevention and care), information construction changes (moving away from patient's data to other related family factors) [4], changes in mindset (moving away from general care to personal care), and changes in the idea of support and systems are all manifestations of this shift (from prescription to preventive medicine). The mantra is to act smart and move with agility toward a smart society 5.0 [16].

These adaptations place an emphasis on catering to the specific requirements of individuals while also boosting the effectiveness of healthcare coverage [10]. As a result, the overall experience of receiving medical and health services is significantly improved, and they point to the direction in which modern medicine will continue to develop in the future [8].

## 9.2 WHAT IS TECHNOLOGY-ENABLED SMART HEALTHCARE?

The term "smart infrastructure" refers to intelligent infrastructure that uses sensors that can detect information and relays information over the network. Supercomputers and Cloud are increasingly being used for this purpose. They are capable of coordinating and integrating many social systems, which enables to accomplish the dynamic and sophisticated administration of human civilization [14]. The trend is toward devices that can be worn-like accessories, the Internet of Things (IoT), and smartphones to monitors, connecting family and support systems who can actively respond smartly in time is referred to as "smart healthcare" [12].

To ensure that the needy are receiving the services they require, ensure stakeholders are acting responsibly based on accurate information, and facilitate the ability to allocate resources efficiently; smart healthcare has the potential to encourage interaction between all stakeholders in healthcare [13]. It is therefore an advanced level of information building in the realm of medical treatment using technology to the best of its ability [11].

As a part of this exploratory study, around 460 experienced doctors (ranging from 10 to 40 years) from various parts of the country were interviewed both objectively and subjectively. Their identity was not disclosed and because of which they gave their opinions freely without any inhibition.

## 9.3 THE ENABLING TECHNOLOGIES

Multiple stakeholders, including medical professionals and patients, healthcare facilities, and academic research organizations, are involved in the delivery of cognitive medical care. It is an organic totality that incorporates numerous dimensions, such as the prevention and monitoring of diseases, the diagnosis and treatment of illnesses, the administration of hospitals, the formulation of decisions regarding healthcare, and the conducting of medical research. Ref. [3] advocated in their research that together with contemporary biotechnology, there are others that also form the foundation of smart healthcare for example, IoT, cloud computing, 5G, artificial intelligence (AI), and so on. Liu et al. [1] advocated that these technologies are utilized extensively across the board in relation to smart healthcare. The concept of "smart aging," that is, the patient is able to wear a device like an accessory which continuously monitors the vital statistics and if there is any deviation, it alerts to seek medical assistance through virtual assistants, and remote services can be taken advantage of within no time at all [15]. A doctor sees the same as a smart support system which improves their diagnosis and prescription. They get the vital figures within no time at all and even if it might not be very accurate, it still gives them a time advantage. An integrated information platform allows for the regular record of patient's data by medical professionals. Ninety percent doctors agreed that using surgical robots and mixed reality technologies, more accurate surgery is now possible. Forty-eight percent respondents report that radiofrequency identification technology is also being effectively utilized to manage equipment and supply chain in hospitals. Thirty percent doctors suggested that this could be better accomplished by utilizing technology to collect data and subsequent analysis. About 60% doctors shared their experience of the utilization of mobile medical platforms which is now resulting in improved experiences for patients. They felt that it is now feasible for scientific research institutes to employ technologies such as AI or deep learning instead of manually diagnosing, prescribing drugs and supplying them to patient in time. About 56% doctors declared this as one of the crucial advantages of adopting these methods. Medical treatments and related hazards may be minimized, medical resources can be better leveraged, exchanges and collaboration across diverse regions can be facilitated, and self-service medical care and medicine through telecommunication can be facilitated using smart healthcare options. These benefits are now being effectively achieved through the utilization of these technologies in every aspect of healthcare shared by 47% respondents of this survey.

The diagnosis and treatment of illnesses have grown more sophisticated because of the application of technologies such as AI, surgical robots, hybrid AI systems, and virtual reality/augmented reality. By constructing the clinical decision support system with the help of AI, it has been possible to obtain several achievements, including the accurate detection of many rare types of malignant cancer. About 76% experienced doctors agreed that the accuracy of diagnoses produced by AI is superior to that produced by human physicians in terms of time, accuracy, and effort. They were all in favor of new technologies but in a regulated and continuously monitored manner (95%).

Systems that are based on machine learning are frequently highly specialized particularly in the fields of diagnosis and radio imaging. IBM's Watson exemplifies the field of medical decision support technique and is widely considered to be its most outstanding and representative example of smart healthcare system. The program has a significant impact on the accuracy of the diagnosis of both cancer and diabetes. About 65% doctors thought it was possible for them to provide patients with expert advice that is based on algorithms through smart healthcare devices. The diagnosis by medical experts was found to be accurate, missed diagnosis and misdiagnosis were rare, and timely prescription and cure became a reality. The state of the patient and the status of the disease may be more correctly characterized based on smart diagnosis, which assists in the development of a specific treatment plan, and the software has been validated by specialists. The actual procedure of therapy will end up being carried out with more accuracy. With the assistance of smart radiomics, for instance, the radiation procedure administered to a patient undergoing treatment for tumors may be dynamically monitored all the way through treatment. The radiation program can be optimized, the illness progression can be monitored, and the uncertainty associated with manual operation may be avoided. The development of surgical robots has made it possible to do surgery at a higher level than ever before. Patients report better results and quicker recovery times as compared to traditional endoscopic surgery, and doctors have access to technology that provides them with greater flexibility and compatibility than traditional endoscopic surgery. The utilization of remote surgery will also result in greater levels of convenience (84%).

The utilization of technology that uses mixed realities makes the process of developing and carrying out the surgical plan more straightforward. Dr. Rajesh Wirth, a senior doctor of a reputed hospital in India reports that the virtual world, the real world, and the users of the system create a circle through matching results with the existing cases. He foresees that the development of this technology further will bring about revolutionary shifts in clinical practice, as well as medical education and research, as well as communication.

Chronic illnesses that are incurable, perhaps terminal, and expensive to treat require smart healthcare options. On the other hand, the conventional paradigm of health management, which is concentrated on hospitals and physicians, is unable to appropriately deal with massive data volumes. About 37% doctors responded favorably on the health management model that is now being implemented as part of smart healthcare and placed a greater emphasis on emerging patient autonomy in managing their own treatment. They reported that patients can themselves monitor and communicate instant feedback of health condition to doctor, leading to timely intervention and diagnosis. The development of smart gadgets that can be implanted or worn, as well as smart homes and health information platforms, all of which are connected by IoT technology, offers a solution to this problem (94%). Wearable and implantable devices combine advanced sensitive sensors, chips, leads, and wireless networks to regularly sense and monitor various physiological indicators of patients from time to time. These devices also can reduce mental stress, give confidence to patients, and interact with other data available from diverse sources to reach expert conclusions. About 65% doctors suggested a transition to frequent monitoring and control procedures to obtain accurate diagnosis. The experience was that it lowered the dangers that are linked with diagnosis because of the sickness. About 56% shared that the proliferation of connected devices such as smartphones, smartwatches, and so on has created a new platform upon which this type of monitoring may take place. There have been several attempts to include biosensors in mobile phones. The mobility of high-performance smartphones continues to improve, and users now have an easier time monitoring both their surroundings and their bodies with these devices (83%).

Those who are elderly or disabled can receive support at home using smart homes. These are specialized homes or flats that monitor the tenants' bodily indications and surroundings using sensors and actuators that are incorporated into the residential infrastructure of the building. The activities that can be carried out also contribute to an enhanced quality of life. About 54% doctors suggested that home automation and health monitoring are the two primary facets that make up the function that smart homes play in the healthcare industry. People who require care are now able to become less reliant on healthcare professionals while simultaneously improving the quality of their lives at home thanks to the ability of these technologies to perform some basic services while simultaneously gathering health data (83%).

Apps and a health information platform enable patients to take charge of their own healthcare and self-manage their conditions.

The traditional method of determining the likelihood of an individual contracting a disease is collecting relevant data, comparing the information with the regulatory state health guidelines, and then diagnosing accordingly. This approach is time-consuming, requires monitoring constantly, and fails to give accurate and precise findings. In the context of smart healthcare, disease risk prediction is reported to be both dynamic and individualized. Around 67% said that it makes it possible for patients and their doctors to take part, to proactively monitor their illness risk, and to engage in preventative efforts that are tailored according to the findings of their own monitoring. The latest model for predicting disease risk collects data using wearable watches (e.g., Apple, Mi) and through interconnected apps, data are stored onto the cloud, and then results are analyzed using neural networks and algorithms (72%).

Afterward, this model utilizes a short messaging platform to transmit the findings of its projections to the participants in perfect sync. The use of these safety precautions has a track record of producing favorable results. They provide support to both physicians and patients in changing their medical practices and lifestyles at any given point in time, and they also provide guidance to decision-makers in the establishment of regional healthcare programs with the intention of lowering the probability of ailment.

Algorithms forming part of these technologies constitute a virtual assistant rather than a tangible person or thing. When interacting with users, robo-advisors employ techniques such as voice activation or sensors relying on massive volumes of information to aggregate information sources, and then provide reactions that are personalized to the user's preferences or needs after performing computations. About 36% doctors were of the opinion that virtual assistants for example the following: Apple Siri, Google Assistant, and Microsoft Cortana may assist patients with a variety of activities, from setting reminders to automating their homes, which make use of session history and language-understanding technologies. The primary function those virtual assistants take on in connected healthcare is that of a conduit via which patients, physicians, and medical organizations may communicate with one another. They (36%) agreed that it would improve the efficiency of receiving medical care. Through the patient's smart device, the virtual assistant could quickly and simply transform common, daily language into one employing medical terminology. The patients can pick and choose the services they want. The virtual assistant can use relevant information based on the patient's past data records. This assists doctors in managing patients and coordinating medical operations in a more convenient manner, which allows doctors to save more time. About 93% doctors suggested the use of virtual assistants in medical facilities that can result in significant savings in terms of both labor and material resources, as well as in improved responsiveness to the requirements of all involved. They (93%) agreed to the possibility to employ new technologies to enable communication between various virtual assistants, which will result in a significant enhancement of the overall experience provided to those who use medical services. Virtual assistants are also assisting in terms of mental health. These assistants can be employed to improve the health of animals or pets too. The patients find these virtual assistants as major stress busters.

Smart healthcare is used throughout biopharmaceutical enterprises for the sake of prescription drug production and dissemination, recording accounts, asset tracking, supply chain management and separating the fake from real, as well as for other tasks that are analogous. Smart healthcare was developed in the United States. About 84% doctors reported here that using RFID technology, it is now possible to provide everyone with their own unique RFID tag, easily accessible through the use of handheld devices. This activity helped to ensure that hospital supplies are moved in a manner that is safe, reliable, stable, and effective. About 78% doctors felt that an integrated management platform can achieve tasks including service provision, quality assessment, evaluation metrics, optimization techniques and performance evaluation. Medical expenses are minimized to an extent, the usage of resources is economic, expert solutions are long-term sustainable and specialized medical institutions are able to take development decisions. In terms of the experience that patients have, they have access to many tasks, such as systems for physical examinations, online appointment scheduling, and confidential communications between doctors and patients. About 60% of doctors interviewed here report that the medical treatment procedures for patients are streamlined because of these automated solutions. About 89% doctors were of the opinion that patients now waited for less time and received service that is more personalized. In a nutshell, majority (95%) doctors surveyed agreed integration, more refining, and further automation are the three primary priorities for the future of smart hospitals and smart healthcare.

## 9.4 ANALYSIS AND THE ROAD AHEAD

The advent of intelligent healthcare has resulted in the development of fully developed ideas and structures. Despite this, there is still a significant amount of potential for improvement, and many new difficulties are currently coming into play because of the introduction of new technologies and new problems. Now, there is a lack of macrodirection and programmatic guidelines in the field of smart healthcare. This results in ambiguous development goals, which eventually leads to a waste of resources. In addition, medical institutions do not have similar standards across different locations and different organizations, and there is a need for changes in the process of ensuring that data integrity is maintained [2]. Because of the sheer volume and complexity of the data, both the exchange of information and the transmission of messages can be challenging. Compatibility issues between the various operating systems and hardware platforms are another source of frustration. The patient feels smart healthcare is lacking in applicable legal norms, and there are dangers around the disclosure of personal information and breaches of privacy [5]. Even for some experienced users, these technologies might be challenging. Some technological advances linked with astute healthcare are still in the exploratory stages from a technological perspective and they ask for a significant amount of financing to be maintained and updated. If it is used carelessly, there is also a possibility of undiscovered danger [7].

Therefore, if we wish to receive answers to the previously identified challenges, we must focus on two distinct factors: regulations and innovative thinking. For a start, the process of associated technologies reaching maturity and stability may gain impetus by implementing updates. Improving one's capacity to derive meaning from massive amounts of data is likewise an essential goal. The establishment of a common technological standard as a means of achieving the highest possible level of interoperability between various platforms and devices is the second key step. It will be possible to increase data integrity and remove obstacles to the sharing of information using this method. Lastly, the utilization of methods such as blockchain is going to be necessary to guarantee the highest level of data security and transmission stability. Regarding the issue of regulation, experts from the many sectors that are pertinent to the topic might collaborate to define the growth goals of the industry. Legislation is necessary for ensuring the confidentiality of patients and enhancing the safety of intelligent healthcare delivery systems [18].

In conclusion, the possibilities for intelligent healthcare are quite promising. Better health self-management is something that can be facilitated for individual users of smart healthcare. When they are required, patients will have access to timely and appropriate medical care, and the services themselves will have a greater degree of personalization. Smart healthcare has the capability to lower the cost for healthcare facilities, alleviate the burden of manpower, enable streamlined organization of data and services, and enhance the experience of patients. Smart healthcare has the potential to lower the costs of conducting research, shorten the amount of time spent conducting research, and increase the overall productivity of research institutes. Smart healthcare may alleviate the conventional wisdom of medical wealth disparities, push medical transformation, encourage the implementation of preventive techniques, and diminish community medical discretionary spending. These are all aspects of macro decision-making that can be affected by smart healthcare [6]. Despite this, there are still some issues with the way the development process is being handled. The answer to these issues will not only be found via the development of new technologies, but also through the collaborative efforts of patients, medical professionals, hospitals, and corporations that specialize in developing new technologies and moving toward a smart society 5.0.

## KEYWORDS


- **intelligent computerization of healthcare**
- **administration of healthcare**
- **IoT**
- **medical devices**
- **smart aging**
- **healthcare technology**


## REFERENCES


1. Liu BH, He KL, Zhi, G. The impact of big data and artificial intelligence on the future medical model. *Journal of Life and Environmental Sciences (PeerJ)* 2018;39:1–4.
2. Martin JL, Varilly H, Cohn J, Wightwick GR. Preface: Technologies for a smarter planet. *IBM Journal of Research and Development* 2010;54(4):1–2.
3. Gong FF, Sun XZ, Lin J, Gu XD. Primary exploration in establishment of China's intelligent medical treatment. *Modern Hospital Management* 2013;11(2):28–29.
4. Farahani B, Firouzi F, Chang V, Badaroglu M, Constant N, Mankodiya K. Towards fog-driven IoT eHealth: Promises and challenges of IoT in medicine and healthcare. *Future Generation Computer Systems* 2018;78:659–676.
5. Wang S, Summers RM. Machine learning and radiology. *Medical Image Analysis* 2012;16(5):933–951.
6. Senbekov M, Saliev T, Bukeyeva Z, Almabayeva A, Zhanaliyeva M, Aitenova N, ... Fakhradiyev I. The recent progress and applications of digital technologies in healthcare: A review. *International Journal of Telemedicine and Applications* 2020;2020:8830200.
7. Ter-Akopov GN, Kosinova NN, Knyazev SA. Digital technologies in healthcare: Achievements and prospects. In *1st International Scientific Conference "Modern Management Trends and the Digital Economy: From Regional Development to Global Economic Growth"(MTDE 2019)*. Atlantis Press, 2019, pp. 321–325.

8. Laurenza E, Quintano M, Schiavone F, Vrontis, D. The effect of digital technologies adoption in healthcare industry: A case based analysis. *Business Process Management Journal* 2018;24:1124–1144.
9. da Rosa VM, Saurin TA, Tortorella GL, Fogliatto FS, Tonetto LM, Samson, D. Digital technologies: An exploratory study of their role in the resilience of healthcare services. *Applied Ergonomics* 2021;97:103517.
10. Wang Q, Su M, Zhang M, Li, R. Integrating digital technologies and public health to fight Covid-19 pandemic: Key technologies, applications, challenges and outlook of digital healthcare. *International Journal of Environmental Research and Public Health* 2021;18(11):6053.
11. Tortorella GL, Saurin TA, Fogliatto FS, Rosa VM, Tonetto LM, Magrabi, F. Impacts of Healthcare 4.0 digital technologies on the resilience of hospitals. *Technological Forecasting and Social Change* 2021;166:120666.
12. Locsin RC, Soriano GP, Juntasopeepun P, Kunaviktikul W, Evangelista LS. Social transformation and social isolation of older adults: Digital technologies, nursing, healthcare. *Collegian* 2021;28(5):551–558.
13. Osipov VS, Skryl TV. Impact of digital technologies on the efficiency of healthcare delivery. In *IoT in Healthcare and Ambient Assisted Living*. Springer: Singapore, 2021, pp. 243–261.
14. Haggerty, E. Healthcare and digital transformation. *Network Security* 2017; 2017(8): 7–11.
15. Raimo N, De Turi I, Albergo F, Vitolla, F. The drivers of the digital transformation in the healthcare industry: An empirical analysis in Italian hospitals. *Technovation* 2022;121:102558.
16. Jadhav VD, Moosafintavida DS. Blockchain in healthcare industry and its application and impact on Covid 19 digital technology transformation. *Mukt Shabd Journal* 2020;9:479.
17. Hermes S, Riasanow T, Clemons EK, Böhm M, Krcmar H. The digital transformation of the healthcare industry: Exploring the rise of emerging platform ecosystems and their influence on the role of patients. *Business Research* 2020;13(3):1033–1069.
18. Wu DD, Olson DL. Assessment of smart healthcare services. In *Pandemic Risk Management in Operations and Finance*. Springer: Cham, 2020, pp. 101–121.

## CHAPTER 10

# ELM-HC: An Approach for Heartbeat Classification Based on Machine–Human Interaction Model

VIPUL NARAYAN[1*], PAWAN KUMAR MALL[1], SWAPNITA SRIVASTAVA[1], PALLAVI JAIN[2], VIMAL KUMAR[1], ANJEY MANI TRIPATHI[2], and SUDESHNA CHAKRABORTY[2]

*[1]Department of Computer Science and Engineering, Madan Mohan Malaviya University of Technology, Gorakhpur, Uttar Pradesh, India*

*[2]Department of Computer Science and Engineering, Galgotias University, Greater Noida, Uttar Pradesh, India*

*Corresponding author

## ABSTRACT

Brain–computer interface is the term given to the process of converting human intents into patterns that may then be used to control equipment that is linked to the outside of the body. An electrocardiogram (ECG) is a visual representation of electrical patterns generated by the heart. ECG is a significant diagnostic tool widely used for people with cardiac issue. ECGs are often recorded on the chapter with a grid on them. The grid work on ECG recordings is made up of numerous small 1 mm square blocks. Each of these little boxes lasts around 40 ms. The transthoracic interpretation of electrical heart activity over time is known as ECG. Assessment of the ECG signal signifies the heart health. For the diagnosis of heart disorders, a number of techniques have been used, including fast Fourier transforms and wavelet transform. In this chapter, the combination of extreme learning machine (ELM) with discrete wavelet

Machine Learning Applications for Data Analysis in Healthcare Systems: Sudeshna Chakraborty, Jyotsna Singh, Praveen Kumar Shukla, Prasenjit Chatterjee (Eds.)


DOI: 10.1201/9781779643193-10

transforms results in what is known as an ELM-healthcare model. This model is primarily made up of four components, and those components are preprocessing, heartbeat segmentation, extraction features, and heartbeat classification. Our proposed model outperforms other state-of-art models and achieved 99.9% accuracy, 98.9% sensitivity, and 99.9% specificity.

## 10.1 INTRODUCTION

Electrocardiogram (ECG) measures heart activity in the electrical form. It is a visual representation of the biopotential's variation throughout time. Leads are attached to certain parts in body to record (ECG) on the graph of the chapter or a computer monitor, depending on the application. Human heart has four chambers: right, left atrium, left, and right ventricular [1, 2]. The upper two chambers are atria, and the bottom two are ventricles. The sino atria node in the right atrium, where the heartbeat normally starts, is responsible for transmitting electrical signals across the heart. The atrio-ventricular (AV) node receives this signal as it travels from the atria [3–5]. The lower chamber's ventricles, which contain a network of fibers that channel electricity and convey impulses to all of its parts, are connected to the AV node. This path of propagation must be accurately traced to make sure the heart is working right. Figure 10.1 depicts an overview of the ECG classification process.

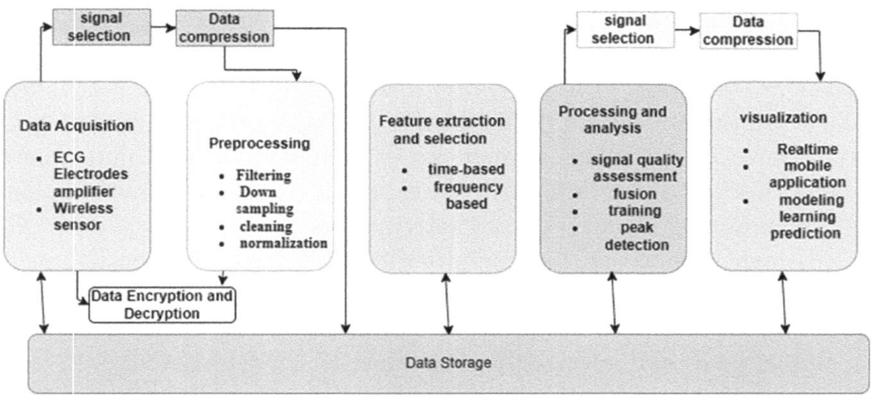


**FIGURE 10.1** Overview of ECG classification process.

### 10.1.1 ECG WAVEFORM

The display of each heartbeat is made up of electrical pulses that have peaks as well as valleys at regular intervals. The length of the electrical wave that passes through the heart affects electrical activity (Figure 10.2). The second element is the level of electrical activity the heart muscle is enduring, which can be used to identify any heart muscle portions that are excessively large or overworked. ECG signals range from 0.05 to 100 Hz. P, Q, R, S, and T represent the ECG's five peaks and valleys for the formation of the signal. There is also sometimes a U wave. The T and P waves, along with the QRS complex, are what make ECG analysis most effective that can be found accurately and reliably [7–9]. The two atria, the heart's upper chambers, are activated by the P wave, while the ventricles. The QRS complex and T wave stimulate the lower chamber of the heart. One of the basic problems with automatic ECG signal analysis is QRS recognition. After detecting QRS complex, ECG signal examined thoroughly. The P, QRS, and T waves show atrial and ventricular depolarization and repolarization [8,10,11]. The baseline is a potential line that refers to horizontal portion of waveform that appears before P wave. Depolarization of the atrial musculature causes the P wave. QRS complex results from simultaneous AV repolarization and ventricle depolarization. T-waves occur during ventricular repolarization, while U waves are assumed to be after potentials in the ventricular muscle. The T wave occurs during ventricular repolarization [12–15]. It is possible to detect cardiac arrhythmias, myocardial infections, ventricular hypertrophy, conduction abnormalities, conduction abnormalities, and other disease states based on the length, amplitude, and form of the QRS complex [16]. The heart beats 60–100 times per minute on average. Slower-than-normal heart rate is bradycardia; faster-than-normal is tachycardia (fast heart) (Table 10.1).

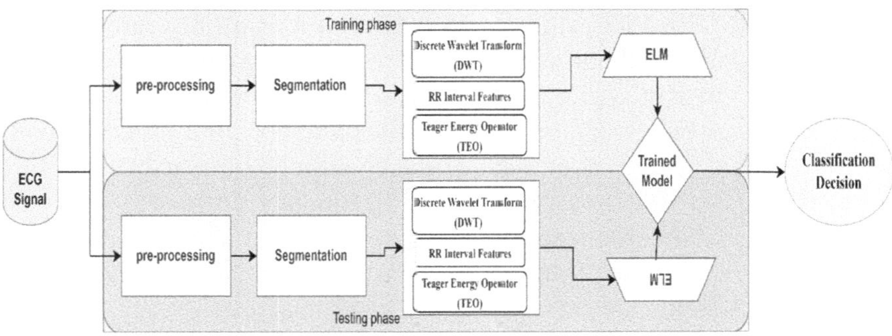


**FIGURE 10.2** ELM-healthcare model workflow diagram.

**TABLE 10.1** Diseases and P Wave

| Dextrocardia | Inverted P wave |
|---|---|
| Hyperkalemia | Tall T-wave without a P wave |
| Sudden cardiac death | Asymmetrical ECG |
| Sinoatrial block | Total interruption of a cardiac cycle |
| Myocardial ischemia | Inverted T wave |

## 10.2 LITERATURE REVIEW

DeepSleepNet is a deep learning model for automatic sleep stage rating based on raw single-channel EEG [17]. The majority of the currently used techniques rely on hand-engineered components, which necessitate prior sleep analysis expertise. Few of them incorporate within the retrieved features the temporal information necessary for determining the upcoming stages of sleep, such as transition rules. In this new model, convolutional neural networks extract time-invariant characteristics and bidirectional long short-term memory learns sleep-stage transition rules from EEG epochs. In this study [18], a method for choosing mother wavelet basis functions is presented. It is used to denoise the ECG signal in the wavelet domain while preserving peak signal amplitudes that are close to their full values. The wavelet-based denoised ECG readings maintain the original diagnostic information [19]. Graph-based attention model uses an attention technique to describe a medical idea as a composite of its ontology ancestors based on the data volume and ontology structure. The classification of ECG beats using a mixed neural network model is demonstrated in this research [20]. ECG signals were dissected using discrete wavelet transform (DWT) and statistical characteristics were derived to describe distribution. In Ref. [21], authors used both the original and noise-reduced sets of ECG signals from a public database for our experiment. The number of instances of each of the five kinds of heartbeats in this set was artificially increased, and high-frequency noise was removed by filtering. CNN was trained utilizing enhanced data and obtained 94.03% and 93.47% accuracy in diagnosing heartbeats in original and noise-free ECGs.

## 10.3 METHODS

### 10.3.1 ECG ANALYSIS

ECG signal analysis and identification has long been a fascinating subject, and there is still ongoing study in this area. For ECG analysis, several fast and short Fourier transform approaches have been developed [22].

#### 10.3.1.1 FAST FOURIER TRANSFORM

Time domain approach was previously employed for ECG signal analysis. However, the drawback was that it was not enough to analyze every aspect of the ECG signal [23, 24]. As a result, the fast Fourier transform (FFT) method was introduced. Fourier transform translates time-domain signals to the frequency domain to find frequency coefficients [25]. FFT is a fundamental transform in digital signal processing that can be applied to a variety of tasks, including frequency analysis and signal processing [26]. It is a faster and more powerful technique that produces the same result as the discrete Fourier transform [27]. The following formula, which can be found in Equation 10.1, is used to define FFT:

$$Xk = \sum N-1\ xe-nk2\pi i/n, n = 0 \tag{10.1}$$

Various characteristics of ECG wave patterns are presented, as well as the condition associated with each type [22] (Table 10.2).

**TABLE 10.2** Abnormalities

| Abnormality | Characteristics |
|---|---|
| Bradycardia | R–R interval is greater than 1 s |
| Tachycardia | R–R interval less than 0.6 s |
| Hypercalcemia | QRS interval less than 0.1 s |

where $k$ is an integer between 0 and $N - 1$ where several methods are used to compress ECG signals.

One of the most important methods is FFT. The following steps make up the entire process:

- Obtaining an input signal or ECG sample.
- Before compression can be performed, the low-frequency components of the incoming signal are filtered out.
- Inverse FFT is used to recover the original signal.

However, the FFT has a limitation in that it cannot provide information about the precise timing of frequency components. This is a disadvantage of the FFT.

#### *10.3.1.2 SHORT-TIME FOURIER TRANSFORM (STFT)*

The windowed-Fourier transform, commonly known as the short-time Fourier transform (STFT), was first introduced by Dennis Gabor in 1946 and thereafter referred to as the Gabor transform [28]. This was developed so that FFT could overcome this limitation. Time and frequency information is contained in STFT. It determines the signal's sinusoidal frequency and time-varying phase content. In comparison to other methods of time–frequency analysis, the STFT-based spectrogram is an easy and efficient process. It is a simple method for slicing the desired waveform into different short segments. Then, it uses the conventional Fourier transform to analyze each segment. A window function is used on a block of information to separate it from the rest of the waveform, and then the Fourier transform is used on that piece of data.

This can be represented with a spectrogram or the STFT. The definition of STFT for a signal $x(t)$ can be found in Equation 10.2 as follows:

$$X(r,\_f_{/2}) = \int \frac{T}{2} x(t)w(t - re - 2\pi ift)dt \tag{10.2}$$

where characteristics of high frequency are represented by $T$. Contrary to STFT, continuous wavelet transform (CWT) is not restricted to sinusoidal analytical functions. It is possible to select localized waveforms that adhere to mathematical standards [29].

Signal $x(t)$ CWT is provided by Equation 10.3. The STFT, where $w(t)$ is a window with duration $T$ and is centered at time point $t$, is the Fourier transform of the windowed signal $x(t)w(t - \tau)$ [16].

$$W(a,b) = \frac{1}{\sqrt{a}} \int \infty f(t) h \times (\frac{t-b}{a}) dt \tag{10.2}$$

STFT's time-frequency accuracy is not ideal. To overcome this, a better technique is used.

#### 10.3.1.3 WAVELET TRANSFORM

The $y$-axis scaling parameter $a$, and the $x$-axis shift parameter $b$ are all related to the mother wavelet, $h(t)$. To explain what the DWT is:

The size of the STFT window:

$$W(j, k) = \sum j \sum k x(k) e(n - k) \tag{10.4}$$

does not give information about the signal at different resolutions. Wavelet transform, on the other hand, has a feature called "multiresolution," which gives information about both time and frequency by letting the window size change [30].

French geophysicist Jean Morlet introduced the Wavelet in 1982.

The term "Wavelet" refers to a little wave, and research into the wavelet transform is a new method for analyzing seismic signals.

Alex Grossmann theoreticians studied the wavelet transform inverse formula [28].

An analytical tool for transitory, nonstationary, or time-varying signals, a wavelet is a tiny wave with energy concentrated in time [31].

There are many different wavelets that can be used in a wide range of situations. Some things about them that make them helpful are as follows:

- Wavelets have a confined frequency and temporal domain.
- Nonstationary signals like ECG exhibit frequent level changes and uneven characteristics.
- Wavelets split signals into multiresolution components.

A time-scale representation known as the wavelet transform has proven useful in several applications, most notably signal compression. The signal is divided into several scales related to frequency components, and each scale is subjected to a linear process that examines it with a specific resolution [32]. Another benefit of the wavelet technique is the availability of a variety of wavelet functions, which allows choosing the optimal function for analyzing the signal as opposed to Fourier analysis, which is limited to one feature morphology, the sinusoid.

#### 10.3.1.4 TEAGER ENERGY OPERATOR (TEO)

Independent RR interval analysis does not capture nonlinear RR interval inconsistency. Teager energy operator (TEO) represented the nonlinear behavior of the energy-tracking RR interval. The signal-generating device's energy, amplitude envelope, and frequency are all measured. The amount of ECG energy required to produce signals with various frequencies and the same energy and amplitude might change. High-frequency signals use more energy. TEO responds to source energy, showing conduction channel instabilities and Teager energy impulses.

#### *10.3.1.5 RR INTERVAL FEATURES*

RR intervals R is the ECG waveform's peak, and RR is the interval between QRS complexes. Nonlinear ECG components vary more than linear equivalents during arrhythmia. RR intervals are easy to calculate and noise resistant. To define the heartbeat's dynamic qualities, previous RR, average RR, post-RR, and local-RR intervals were retrieved from the RR sequence.

#### *10.3.1.6 EXTREME LEARNING MACHINE (ELM)*

Extreme learning machine (ELM) is a feedforward neural network with one hidden layer and three layers: input, hidden, and output. Figure 10.3 shows ELM's structure. ELM uses different parameters than blood pressure. The parameters for the hidden layer are produced at random, whereas the parameters for the output layer are computed analytically by applying least-squares regression. ELM approximates across the board. A network with hidden layer parameters that are generated at random has the ability to approximate any restricted noncontact piece-wise continuous activation function with an error that can be arbitrarily small.

#### *10.3.1.7 DATASET*

This study uses data from the PhysioNet MIT-BIH Arrhythmia and PTB Diagnostic ECG Databases [33]. We also demonstrate how data from the first database may be successfully moved to the second to train inference models. In all of our tests, the input was ECG lead II resampled to a sampling frequency of 125 Hz. The ECG recordings from 47 different individuals

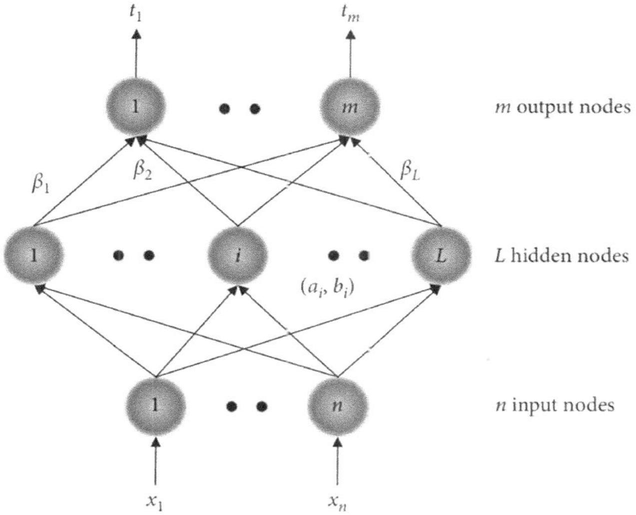


**FIGURE 10.3** ELM block diagram.

were recorded at a sampling rate of 360 Hz for the MIT-BIH dataset. At least two cardiologists annotate each pulse. In accordance with the Association for the Advancement of Medical Instrumentation EC57 guideline, we use annotations in this dataset to define five distinct beat types [34].

Table 10.2 displays beat annotation mappings by category. The PTB Diagnostics dataset contains ECG recordings from 290 people: 148 with myocardial infarction, 52 healthy controls, and the rest with seven different disorders. Each recording has twelve 1000 Hz ECG leads. This study used healthy control groups and ECG lead II (Table 10.3).

### *10.3.1.8 METHODOLOGY*

The proposed model is categorized in four subapproaches such as (1) preprocessing, (2) heartbeat segmentation, (3) feature extraction, and (4) heartbeat classification using ELM.

#### ***10.3.1.8.1 ECG Signal Preprocessing***

Typically, baseline wander or power line interface (PLI) affects ECG signals. To eliminate these forms of disturbances from the ECG signal, many approaches have been proposed [18]. Experiments showed that

**TABLE 10.3** Category-Wise Beat Annotations

| Category | Annotation |
|---|---|
| V | Premature ventricular contraction<br>Ventricular escape |
| N | Atrial escape<br>Normal<br>Nodal escape<br>Left/Right branch block |
| F | Fusion of ventricular and normal |
| S | Superventricular premature<br>Nodal premature<br>Aberrant atrial premature<br>Atrial premature |
| Q | Fusion of paced and normal<br>Paced<br>Unclassifiable |

baseline wander has a big effect on finding arrhythmia and makes it hard for professionals to process ECG signals [19]. PLI is a noise type that happens when you record an ECG signal because of things like broken electrodes, voltage differences between the electrodes, patient movement, breathing mistakes, or resistance between the electrodes. It is a low-frequency noise that usually happens between 0 and 0.3 Hz. To get an ECG signal without noise, take the raw ECG signal and subtract the expected value of the signal from it.

#### *10.3.1.8.2 Heartbeat Segmentation*

The T wave, QRS complex, and P wave are called fiducial points. Correct ECG segmentation includes detecting the fiducial spots' boundaries and peaks. A single pulse included the R peak and 200 surrounding samples. This segment size contained the most information for each heartbeat.

#### *10.3.1.8.3 Feature Extraction*

RR interval, DWT, and Teager energy operator features were chosen because they better represented dynamic and morphological ECG alterations.

#### 10.3.1.8.4 Heartbeat Classification (ELM)

During the feature extraction step, wavelet time frequency and RR interval features are extracted. ELM is trained over 5000 epochs using these features. In the next phase of research work, we will analyze the outcome of the training and testing process.

## 10.4 PERFORMANCE ANALYSIS

The performance analysis is done based on four parameters: TP, TN, FP, and FN which is true positive, true negative, false positive, and false negative, respectively. Equation 10.5 is used to calculate the "Accuracy" as follows.

### 10.4.1 ACCURACY (ACC)

The accuracy (ACC) [35] is calculated as the total TP ($\alpha$) and TN ($\lambda$) records count divided by the total number of instances ($\alpha$ = TP, $\lambda$ = TN, $\beta$ = FN, $\Omega$ = FP). Equation 10.5 is used to calculate the "Accuracy" as follows:

$$\text{ACC} = \frac{\alpha + \lambda}{\alpha + \lambda + \beta + \Omega} \tag{10.5}$$

### 10.4.2 PRECISION (PE)

The precision (PE) [36] is calculated as the ratio of total true positive ($\alpha$) records count by the total number of instances ($\alpha$ = TP, $\beta$ = FN). The "precision" is calculated using Equation 10.6 as follows:

$$\text{SN} = \frac{\alpha}{\alpha + \beta} \tag{10.6}$$

### 10.4.3 RECALL (RE)

The recall (RE) [37] is calculated as the ratio of total true positive ($\alpha$) records count by the total number of true instances ($\alpha$ = TP, $\Omega$ = FP). The "Recall" is calculated using Equation 10.7 as follows:

$$RE = \frac{\alpha}{\alpha + \Omega} \tag{10.7}$$

This model is primarily made up of three components, and those components are noise filtering, noise detection, and classification. Our proposed model outperforms other state-of-art models and achieved 99.9% accuracy, 98.9% sensitivity, and 99.9% specificity (Figure 10.4, Table 10.4).

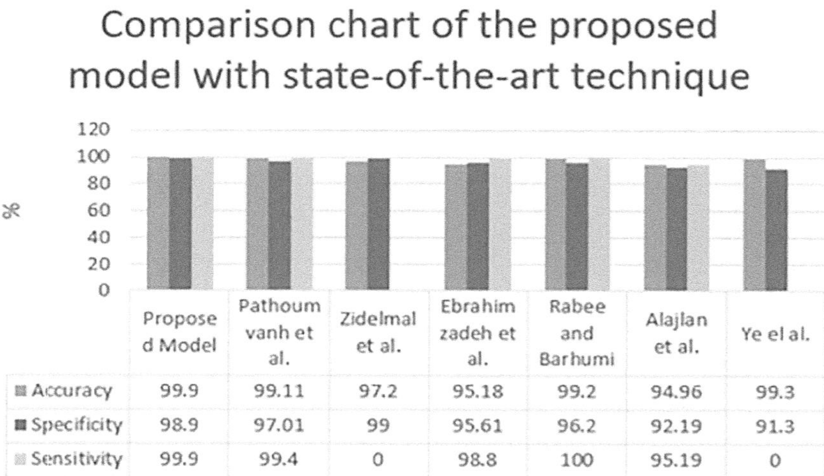


**FIGURE 10.4** Result analysis of the proposed model.

**TABLE 10.4** Proposed Model versus State-of-the-Art Approach Comparison Table

| | Features | Classes | Dimension | Accuracy | Sensitivity | Specificity |
|---|---|---|---|---|---|---|
| Proposed model | DWT + RR + TEO + ELM | 18 | 17 | 99.9 | 98.9 | 99.9 |
| Zidelmal et al. | Frequency content + RR + QRS | 2 | 13 | 97.2 | 99 | – |
| Ebrahimzadeh et al. | HOS + timing interval + DCT | 5 | 24 | 95.18 | 95.61 | 98.8 |
| Rabee et al. | WT multiresolution | 14 | 251 | 99.2 | 96.2 | 100 |
| Alajlan et al. | HOS of second order | 2 | 604 | 94.9 | 92.1 | 95.1 |
| Ye et al. | WT + ICA + RR | 16 | 18 | 99.3 | 91.3 | – |
| Pathoumvanh et al. | DCT | 5 | 5 | 99.11 | 97.01 | 99.4 |

## 10.5 CONCLUSION

An ECG is a graphic representation of the electrical patterns the heart produces. This test is extensively used for persons with cardiac difficulties and is an important diagnostic tool. ECGs are frequently recorded on chapters with a grid. The grid work on ECG recordings is made up of many little 1 mm square blocks. Each of these little boxes lasts around 40 ms. Image processing and numerical data extraction from ECGs. A transthoracic interpretation of the heart's electrical activity over time is called an ECG. The study of the ECG signal provides details regarding the condition of the heart. A variety of approaches, including FFTs and wavelet transforms, have been employed to diagnose cardiac problems. The combination of ELM with discrete wavelet transform resulted in an ELM-healthcare model in this work. This model is basically comprised of four components: preprocessing, heartbeat segmentation, feature extraction, and heartbeat classification. With 99.9% accuracy, our proposed model surpassed existing state-of-the-art models, 98.9% sensitivity, and 99.9% specificity. In the future, this proposed model can be applied to medical imaging issues and different modalities.

## KEYWORDS

- **electrocardiogram**
- **continuous wavelet transform**
- **sino atria**
- **artioventricular**
- **Fourier transform**

## REFERENCES

1. Gautam A, Kaur M. ECG analysis using continuous wavelet transform (CWT). *IOSR Journal of Engineering* 2012;2(4):632–635.
2. Nagendra H, Mukherjee S, Kumar V. Application of wavelet techniques in ECG signal processing: an overview. *International Journal of Engineering, Science and Technology* 2011;3(10):7432–7443.
3. Josko A, Rak RJ. Effective simulation of signals for testing ECG analyzer. *IEEE Transactions on Instrumentation and Measurement* 2005;54(3):1019–1024.

4. Narayan V, Daniel AK, Rai AK. Energy efficient two tier cluster based protocol for wireless sensor network. In: *2020 International Conference on Electrical and Electronics Engineering (ICE3)*, 2020, pp. 574–579.
5. Narayan V, Daniel AK, Chaturvedi P. FGWOA: An efficient heuristic for cluster head selection in WSN using fuzzy based grey wolf optimization algorithm. 2022.
6. Narayana KVL, Rao AB. Wavelet based QRS detection in ECG using MATLAB. *Innovative Systems Design and Engineering* 2011;2(7):60–69.
7. Saritha C, Sukanya V, Murthy YN. ECG signal analysis using wavelet transforms. *Bulgarian Journal of Physics* 2008;35(1):68–77.
8. Narayan V, Daniel AK. A novel approach for cluster head selection using trust function in WSN. *Scalable Computing Practice and Experience* 2021;22(1):1–13.
9. Narayan V, Daniel AK. RBCHS: Region-based cluster head selection protocol in wireless sensor network. In: *Proceedings of Integrated Intelligence Enable Networks and Computing*, Springer, 2021, pp. 863–869.
10. Narayan V, Daniel AK. Design consideration and issues in wireless sensor network deployment. *Invertis Journal of Science & Technology* 2020;101.
11. Pramanik S et al. A novel approach using steganography and cryptography in business intelligence. In: *Integration Challenges for Analytics, Business Intelligence, and Data Mining*, IGI Global, 2021, pp. 192–217.
12. Narayan V, Daniel AK. IOT based sensor monitoring system for smart complex and shopping malls. In: *International Conference on Mobile Networks and Management*, 2021, pp. 344–354.
13. Narayan V, Daniel AK. CHHP: Coverage optimization and hole healing protocol using sleep and wake-up concept for wireless sensor network. *International Journal of System Assurance Engineering and Management* 2022;1–11.
14. Choudhary S, Narayan V, Faiz M, Pramanik S. Fuzzy approach-based stable energy-efficient AODV routing protocol in mobile ad hoc networks. In: *Software Defined Networking for Ad Hoc Networks*, Springer, 2022, pp. 125–139.
15. Narayan V, Daniel AK. CHOP: Maximum coverage optimization and resolve hole healing problem using sleep and wake-up technique for WSN. *ADCAIJ Advances in Distributed Computing and Artificial Intelligence Journal* 2022;11(2):159–178.
16. Supratak A, Dong H, Wu C, Guo Y. DeepSleepNet: A model for automatic sleep stage scoring based on raw single-channel EEG. *IEEE Transactions on Neural Systems and Rehabilitation Engineering* 2017;25(11):1998–2008, https://doi.org/10.1109/TNSRE.2017.2721116.
17. Singh BN, Tiwari AK. Optimal selection of wavelet basis function applied to ECG signal denoising. *Digital Signal Processing* 2006;16(3):275–287, https://doi.org/10.1016/j.dsp.2005.12.003.
18. Choi E, Bahadori MT, Song L, Stewart WF, Sun J. GRAM: Graph-based attention model for healthcare representation learning. In: *Proceedings of the 23rd ACM SIGKDD International Conference on Knowledge Discovery and Data Mining*, 2017, pp. 787–795.
19. Güler İ, Übeylı˙ ED. ECG beat classifier designed by combined neural network model. *Pattern Recognition* 2005;38(2):199–208, https://doi.org/10.1016/j.patcog.2004.06.009.

20. Acharya UR et al. A deep convolutional neural network model to classify heartbeats. *Computers in Biology and Medicine* 2017;89:389–396, https://doi.org/10.1016/j.compbiomed.2017.08.022.
21. Muthuchudar A, Baboo SS. A study of the processes involved in ECG signal analysis. *International Journal of Science and Research* 2013;3(3):1–5.
22. Ranjan R, Giri VK. A unified approach of ECG signal analysis. *International Journal of Soft Computing and Engineering* 2012;2(3).
23. Karpagachelvi S, Arthanari M, Sivakumar M. ECG feature extraction techniques-a survey approach. *arXiv Prepr. arXiv1005.0957*, 2010.
24. Haque AKMF. Improved detection of ECG features using wavelet for emergency medical application. *International Journal of Advance Innovations, Thoughts & Ideas* 2012;1(2).
25. Singh P, Yadav OP, Yadav Y. ECG signal compression implementation by a new 2-dimensional transform technique. *International Journal of Engineering and Advanced Technology* 2012;1(6):167–170.
26. Dangi V, Parab A, Pawar K. Analysis of transform based ECG compression techniques. In: *Proceedings of the International Conference on Advances in Electronics and Electrical Engineering (AEEE 2012)*, 2012, pp. 33–37.
27. Sifuzzaman M, Islam MR, Ali MZ. Application of wavelet transform and its advantages compared to Fourier transform. *Journal of Physical Sciences*. 2009;13:121–134.
28. Addison PS. Wavelet transforms and the ECG: A review. *Physiological Measurement* 2005;26(5):R155.
29. Karthikeyan P, Murugappan M, Yaacob S. ECG signal denoising using wavelet thresholding techniques in human stress assessment. *International Journal on Electrical Engineering and Informatics* 2012;4(2):306.
30. Alfaouri M, Daqrouq K. ECG signal denoising by wavelet transform thresholding. *American Journal of Applied Sciences* 2008;5(3):276–281.
31. Mehta P, Kumari M. QRS complex detection of ECG signal using wavelet transform. *International Journal of Applied Engineering Research* 2012;7(11):1889–1893.
32. Wayne P et al. Tai Chi, Physiological Complexity, and Healthy Aging-Gait. *PhysioNet*. https://doi.org/10.13026/gq9q-rr81
33. Mathunjwa BM, Lin YT, Lin CH, Abbod MF, Shieh JS. ECG arrhythmia classification by using a recurrence plot and convolutional neural network. *Biomedical Signal Processing and Control* 2021;64:102262.
34. Mall PK, Singh PK, Yadav D. GLCM based feature extraction and medical X-RAY image classification using machine learning techniques. In: *2019 IEEE Conference on Information and Communication Technology*, 2019, pp. 1–6.
35. Mall PK, Singh PK. BoostNet: A method to enhance the performance of deep learning model on musculoskeletal radiographs X-ray images. *International Journal of System Assurance Engineering and Management* 2022;1–15.
36. Irfan D, Tang X, Narayan V, Mall PK, Srivastava S, Saravanan V. Prediction of quality food sale in mart using the ai-based TOR method. *Journal of Food Quality* 2022;2022.

# CHAPTER 11

# Bow-Tie Construction from Accident Narratives: A Text-Mining Approach

ASHISH GARG, SOUVIK DAS*, AMARDEEP KUMAR, and J. MAITI

*Indian Institute of Technology, Kharagpur, West Bengal, India*

*Corresponding author

## ABSTRACT

In this chapter, we propose an integrated natural language processing (NLP) and the Bow-tie-based risk assessment method to analyze the accident or incident reports for providing preventive and protective actions to the organization. Over 500 accident/incident text narratives of distinct scenarios of workers at different working sites in an organization were collected. We first applied the custom-trained named entity recognition model and found out the objects involved and the potential root cause of the accident or incident from each of the text narratives. Second, using the elbow method, it is found that 4–6 clusters of root causes can be created. Then, the root causes are classified into four categories namely, human factors, component factors, management factors, and environmental factors, using term frequency–inverse document frequency vectorizer and linear support vector classifier. In order to visualize these accidents/incidents, root causes, and their categories, bow-tie methodology is applied. The preventive and protective barriers have been suggested for each potential top event. Thus, this work provides a qualitative risk assessment method to manage future risk events.

Machine Learning Applications for Data Analysis in Healthcare Systems: Sudeshna Chakraborty, Jyotsna Singh, Praveen Kumar Shukla, Prasenjit Chatterjee (Eds.)


DOI: 10.1201/9781779643193-11

## 11.1 INTRODUCTION

Accidents or incidents are common in manufacturing and other industries. As stated by the International Labor Organization, more than 27.8 million occupational workers die every year due to accidents [10]. One of the major reasons which lead to fatal accidents is the availability of hazardous materials like oil, flammable gas, and so on [6, 9, 11]. These occupational accidents affect the entire organization in different ways; hence stakeholders' safety should be of utmost priority in any industry.

Nowadays in the era of Industry 4.0, industries are collecting ample amounts of data through various means. However, it is difficult to get structured data easily. Also, the value of the data comes from the analysis and insights generated. In manufacturing industries, it is common to find accident or incident reports. These reports are prepared after the accidents by the inspecting agency in conjunction with the witnesses of the accident or incident. Analysis of these accidents or incident reports may provide some great insights or hidden causes of accidents or incidents that may be helpful in reducing the risk in the future. For example, suppose after analyzing accidents reports, it is found that irregular inspection of machines before their day-to-day operation is one of the measure causes of the accidents. These findings may lead to making regular inspection as a compulsory element of standard operating procedures of operations on machines.

The research on the analysis of the accident/incident reports in the manufacturing industries is at its infancy. These researches are mainly focused on natural language processing (NLP)-based tools and techniques. However, integration of quantitative risk assessment techniques such as bow-tie analysis with the NLP is rarely been observed in the literature. Bow-tie is a graphical tool combining fault tree and event tree. A fault tree is drawn from top to bottom and quantitatively analyzed from bottom to top, whereas an event tree is useful in analyzing the risk control measures, that is, safety barriers and the accident consequences. Bow-tie is a highly useful to identify the root causes of any accident and finding the connections between risk control systems. Hence, the motive for this work is to analyze the accident reports with the help of the NLP and the bow-tie analysis. The combination of the results of the text narratives analysis and the bow-tie model provides an improved analysis and better insights. Here, first the preprocessing of the text data such as converting to lowercase, punctuation removal, stop word removal, tokenization, stemming, and lemmatization is done to bring the unstructured text information into a desirable format so that risk assessment can be done easily [7]. To find the root cause of accidents, a custom-trained NER (named entity recognition) model is used. Basically, NER model is a method to extract the key entity from the text. But before applying the NER model, it is necessary to annotate the text data. Over 450 text narratives are annotated. Further, two desired entities, namely, root causes and objects, involved in an accident/incident are labeled. The causation sentences are categorized into four levels by identifying the number of clusters using the elbow method and linear support vector classifier (SVC). It is found the most frequently occurring words in whole corpus, called top event related to accidents are fire, fall down, slipped, and gas leakage. Along with this, the bow-tie diagram is plotted using the open-source BowtieXP software. Bow-tie diagram gives an effective risk assessment process to systematically assess the potential risks.

## 11.2 LITERATURE REVIEW

Nowadays every industry wants to improve its safety performance by reducing the number of accidents or incidents. For this purpose, it is necessary to record and analyze the accidents/incidents/near-misses. In this direction, several researches are conducted in the past decade using the NLP methods. For example [2], presented naïve Bayesian modeling division compensation that includes claims for work-related injuries. The proposed model achieved almost 90% of accuracy, but the accuracy of the applications is limited to the categories of injuries. Taylor et al. [12] used the naïve Bayesian model and fuzzy model to differentiate the effects of the damage. Tixier et al. [13] used stochastic gradient tree boosting algorithms and random forest (RF) to predict the type of root cause, type of injury, affected body part, and severity of injury at the construction site to analyze construction-related risks. Probability Rank Skill Score for risk analysis is rated in the range of 0.236–0.436. Tixier et al. [14] suggested an NLP approach to extract the information from the preliminary reports of random injuries and receive 0.95 accuracies, but the proposed method did not work in unexpected situations. Goh and Ubeynarayana [5] used RF, support vector machine (SVM), *K*-nearest neighbors, linear regression, and naïve Bayes (NB) algorithms including classification of construction hazards. Among them, SVM scored $F_1$ points from 0.45 to 0.92 and was more efficient than other methods. Chokor et al. [3] used the *K*-means method to categorize injury reports. The types of accidents identified were "fall," "hit," "electricity," and "canals falling," respectively. Four sets of injuries were found and each set represented a type of risk. Fan and Li [4] conducted research on how to extract text with a criteria-based consultation process for retrieval of accident records. The result has shown that mining methods are superior in terms of recall and accuracy. Zou et al. [16] proposed an NLP approach with the expansion of the semantic query and vector space model to analyze similar accident cases. An approach for automatic extraction of root cause using semantic similarity has been developed by Cambridge University. To convert the unstructured text data, they followed a preliminary preprocessing such as semantic parsing, morphological analysis, subject–verb object extraction, and NLP-based approach [8]. The steps to draw bow-tie are explained by [1]. The relationship between the root causes, preventive barriers, protective barriers, and consequences are also explained using bow-tie. Bow-tie gives a structured way of visualizing the top events and of consequences accidents/incidents. However, there is dearth of literature where NLP methods are integrated with the traditional quantitative risk assessment techniques such as bow-tie [15]. Hence, in this study, we are trying to integrate NLP and bow-tie methods to analyze text narratives of accident reports. The main objectives are to find the root cause of accident using NLP, categories the accident root cause into four categories: (1) human factor, (2) management factor, (3) environmental factor, and (4) component factor and finally preparation of bow-tie model to visualize these accident causations.

## 11.3 METHODOLOGY

To achieve the above-mentioned objectives, an integrated methodology has been proposed as shown in Figure 11.1.

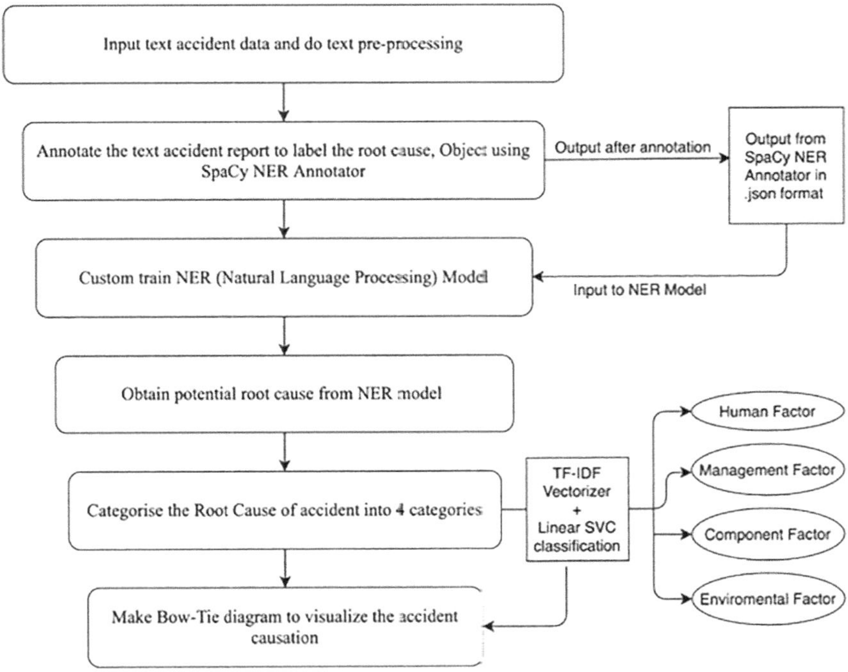


**FIGURE 11.1** Overall methodology.

### 11.3.1 DATA PREPROCESSING

*Stop words removal*: Removal of stop words(.) from sentence. Since these words are not useful for text information, it is to be removed before applying classification model. Some of the common stop words are "a," "an," "the," "is," "as," and so on. In Python, we can easily remove stop words using Natural Language Toolkit (NLTK). NLTK has a list of stop words in their dictionary; thus, we can first convert the whole corpus into list of words and remove the words that match with stop words in the dictionary of NLTK.

*Punctuation removal*: Removal of unnecessary symbols in a document. We remove unknown characters and symbols, such as "#," "!," "$," so that we can find the root words that carry the actual information of the text.

*Convert to lowercase*: Convert to lower or uppercase for comparison ease. We either convert the whole corpus in lowercase or uppercase. In

Python, we can use function lower("string") to convert into lowercase and function upper("string") to convert into uppercase.

*Tokenization*: Split the text into a list of words. Before applying NLP, we first tokenize the text into list of words. We do it because the meaning of the text can be interpreted by analyzing the words.

*Stemming*: Reducing derived word into stem/root word. It is a normalization technique. It helps to find out similar words when we search in our whole corpus.

*Lemmatization*: Similar to Stemming but returns actual dictionary word. Thus, the word we get after lemmatization carries similar meaning to the original word.

### *11.3.2 SPACY NER*

NER model is a pretrained NLP model, which is capable of extracting and categorizing key information from text. The key entities are either words or a series of words and every extracted key entity is classified into a predefined category/label. Output from pretrained NER model is shown in Figure 11.2.

```
import spacy
nlp=spacy.load('en_core_web_sm')

sentence='since April India is experiencing a huge increase in covid19 case majorly in Pune city'

doc=nlp(sentence)

for ent in doc.ents:
    print(ent.text, ent.start_char, ent.end_char, ent.label_)
```

```
April 6 11 DATE
India 12 17 GPE
Pune 77 81 GPE
```

**FIGURE 11.2** Pretrained spacy NER.

### *11.3.3 SPACY NER ANNOTATION*

While using the custom-trained NER model, we first needed the prelabel data. We do annotation so that we can add new label to our model (spacy NER), then later the label can be predicted for the new text/document. The pretrained model can only identify this type of label such as ["DATE," "PERSON," "TIME," "GPE," "PRODUCT," "ORG," "LOC," "MONEY"].

We did annotation (Figure 11.3) first and used the annotated data as an input to the NER model. The annotated data we get from spacy NER annotator is in a json format which is a list of dictionaries. Thus, we find out the root cause using a custom-trained model.

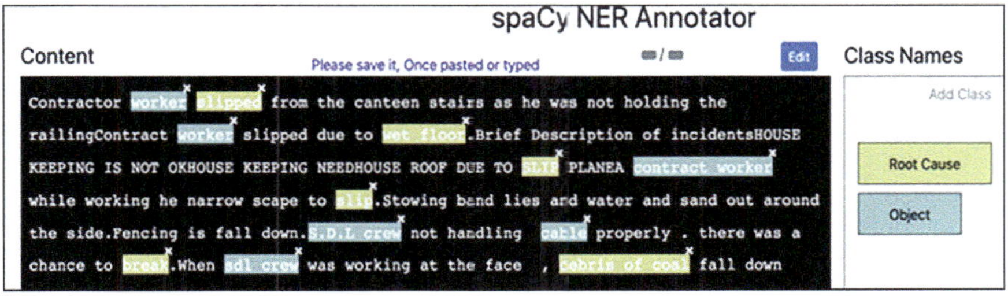


**FIGURE 11.3** Annotation.

### *11.3.4 CALCULATION OF TF–IDF (TERM FREQUENCY–INVERSE DOCUMENT FREQUENCY)*

Term frequency–inverse document frequency (TF–IDF) gives us an idea about which word is most important in our documents. The document which has similar words will carry similar vector which is important for our machine learning model. The highest TF–IDF score represents a rare word in the document. This score thus generates vector of all words and used as an input to classification model such as linear SVC, naïve Bayes (NB) classifier.

*TF*: TF is a way of normalization. The formula for TF is given as follows:

TF($t$) = (number of times word "$t$" appears in a document)/(total number of words in the document)

$$tf_{i,j} = \frac{n_{i,j}}{\sum_k n_{i,j}}$$

*IDF*: IDF measures how important a term is. Basically, it is to scale down the impact of tokens that occur very frequently in every document, such as "a," "the."

IDF($t$) = $\log_e$(total number of documents/number of documents with term $t$ in it).

$$idf(w) = \log\left(\frac{N}{df_t}\right)$$

*TF–IDF*: TF–IDF is simply the TF multiplied by the IDF.

$$w_{i,j} = tf_{i,j} \times \log\left(\frac{N}{df_t}\right)$$

For every accident document, we get the TF–IDF value for each word, and the higher the TF–IDF value signifies the importance of that word. Thus, across each document, we get the most important word by seeing the word containing the highest TF–IDF value in that corresponding document. This TF–IDF matrix is used as an input for the linear SVC.

**Example:** Let us consider the following two sentences:

**Sentence 1:** Coal fell down but no injury. **Sentence 2:** Sdl crew slipped but no injury.

List of unique words: [coal, fell, down, but, no, injury, sdl, crew, slipped]

$$\text{Count of word in sentence-1 } (n1) = 6$$

$$\text{Count of word in sentence-1 } (n1) = 6$$

Term frequency for each word in Sentence-1 is shown in Table 11.1.

TF($t$) = (No. of times word "$t$" appears in a document)/(total number of words in the document)

**TABLE 11.1** Term Frequency in the First Sentence

| Sentence-1 | Coal | Fell | Down | But | No | Injury |
|---|---|---|---|---|---|---|
| TF | 1/6 | 1/6 | 1/6 | 1/6 | 1/6 | 1/6 |

Similarly, term frequency for each word in Sentence-2 is given in Table 11.2.

**TABLE 11.2** Term Frequency in the Second Sentence

| Sentence-2 | sdl | Crew | Slipped | But | No | Injury |
|---|---|---|---|---|---|---|
| TF | 1/6 | 1/6 | 1/6 | 1/6 | 1/6 | 1/6 |

IDF for each word in the whole corpus is given in Table 11.3.

IDF($t$) = $\log_e$(Total number of documents/Number of documents with term $t$ in it).

**TABLE 11.3** Inverse Document Frequency (IDF) for Each Word in Whole Corpus

| Unique Words | IDF |
|---|---|
| Coal | $\log_e(2/1)$ |
| Fell | $\log_e(2/1)$ |
| Down | $\log_e(2/1)$ |
| But | $\log_e(2/2)$ |
| No | $\log_e(2/2)$ |
| Injury | $\log_e(2/2)$ |
| Sdl | $\log_e(2/1)$ |
| Crew | $\log_e(2/1)$ |
| Slipped | $\log_e(2/1)$ |

TF–IDF for each word in Sentence-1 and Sentence-2 is computed as in Table 11.4.

**TABLE 11.4** Computation of TF–IDF for Each Word in Sentence-1 and Sentence-2

| Unique Words | Sentence-1 | Sentence-2 |
|---|---|---|
| Coal | $1/6 \times \log_e(2/1)$ | 0 |
| Fell | $1/6 \times \log_e(2/1)$ | 0 |
| Down | $1/6 \times \log_e(2/1)$ | 0 |
| But | $1/6 \times \log_e(2/2)$ | $1/6 \times \log_e(2/2)$ |
| No | $1/6 \times \log_e(2/2)$ | $1/6 \times \log_e(2/2)$ |
| Injury | $1/6 \times \log_e(2/2)$ | $1/6 \times \log_e(2/2)$ |
| Sdl | 0 | $1/6 \times \log_e(2/1)$ |
| Crew | 0 | $1/6 \times \log_e(2/1)$ |
| Slipped | 0 | $1/6 \times \log_e(2/1)$ |

TF–IDF for each word in Sentence-1 and Sentence-2 is given in Table 11.5.

**TABLE 11.5** TF–IDF for Each Word in Sentence-1 and Sentence-2

| Unique Words | Sentence-1 | Sentence-2 |
|---|---|---|
| Coal | 0.301 | 0 |
| Fell | 0.301 | 0 |
| Down | 0.301 | 0 |
| But | 0 | 0 |
| No | 0 | 0 |
| Injury | 0 | 0 |
| Sdl | 0 | 0.301 |
| Crew | 0 | 0.301 |
| Slipped | 0 | 0.301 |

## 11.4 CASE STUDY

### 11.4.1 DATA SET

We collected over 500 accident text data from the workplace at distinct working sites. Initially, text was in raw format, so first, we did some preprocessing: stop words removal, convert to lowercase and punctuation removal, tokenization, stemming, lemmatization, and brought the text data into a desired usable format. About 450 text data having rows of multiple sentences were manually annotated and stored in json format as an input to the custom-trained NER model.

We prelabeled the text data and then used TF–IDF and linear SVC to classify them into four categories: (1) human factor, (2) component factor, (3) management factor, and (4) environmental factor.

Before this, we used the elbow method to find out the optimal number of clusters that fit for this text data and we got *k* in the range 4–6.

### 11.4.2 DATA VISUALIZATION

Figure 11.4 is the word cloud of the most occurring keywords in sentence after preprocessing such as

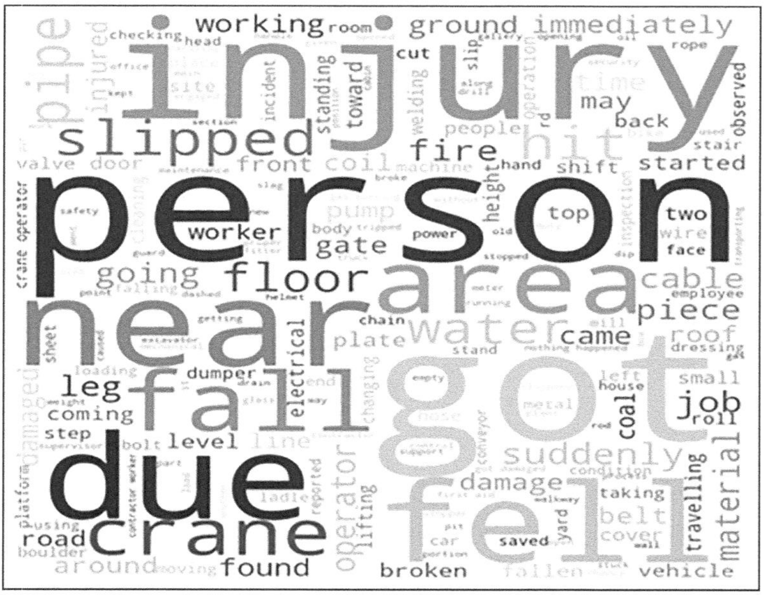


**FIGURE 11.4** Word Cloud.

- Change text to lowercase
- Remove stop words
- Remove bad characters
- Remove punctuation

Overall, it gives an approximate idea about which word is most important. From Figure 11.4, we can extract top events like [slipped, fell, fall, broken, damage, fire].

## 11.5 RESULTS

### *11.5.1 CUSTOM-TRAINED NER MODEL*

We find out the root cause and object using a custom-trained NER model and target sentence using regular expression. Root cause is the factor which may lead to an accident. We find from the NER model that the most frequent root cause words are [fall, fall down, fell, slipped, gas leakage, hydraulic leakage, fire] (Figure 11.5).

| | Ptential_Sentence | Root_Cause | Object | Target_Sentence | Symbol |
|---|---|---|---|---|---|
| 0 | S.D.L crew not handling cable properly . there was a chance to break. | not handling | cable | S.D.L crew not handling cable properly . | X1 |
| 1 | When sdl crew was working at the face , debris of coal fall down near the worker but there was no ijured | fall down | coal | debris of coal fall down near the worker but there was no ijured | X2 |
| 2 | Durin sdl cable handling one sdl crew slipped but not injured. | slipped | cable | Durin sdl cable handling one sdl crew slipped but not injured. | X3 |
| 3 | During pipe extaintion the pipe fitter was fell down but no injury. | fell down | pipe | During pipe extaintion the pipe fitter was fell down but no injury. | X4 |
| 4 | when two sdl crew was going to the face suddenly side loose coal was fall and they were escape nerroly from injured. | fall | coal | when two sdl crew was going to the face suddenly side loose coal was fall and they were escape nerroly from injured. | X5 |

**FIGURE 11.5** Output from custom NER model.

### *11.5.2 CLASSIFICATION*

We categorize the root cause word into four levels of failure, that is human factor, component factor, environment factor, and management factor by using the TF–IDF matrix as an input to the linear SVC. We used 450 data for training and the rest 50 for testing purposes.

We observed that factor that leads to most accidents are management factor and environment error that causes the least. Out of 50 accidents, 28 happened due to management errors (Figure 11.6a–c).

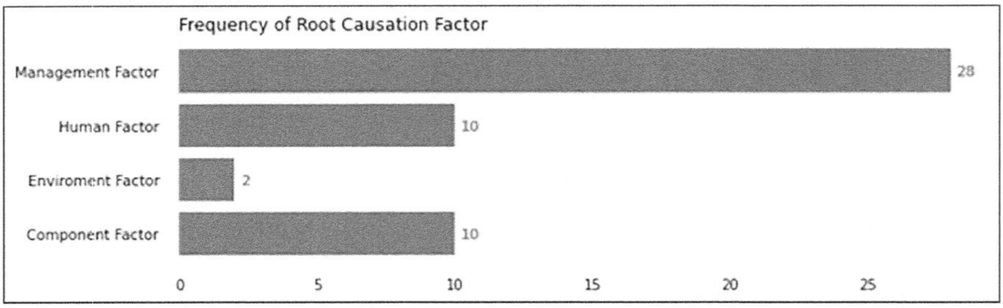


**FIGURE 11.6 (a)** Frequency distribution.

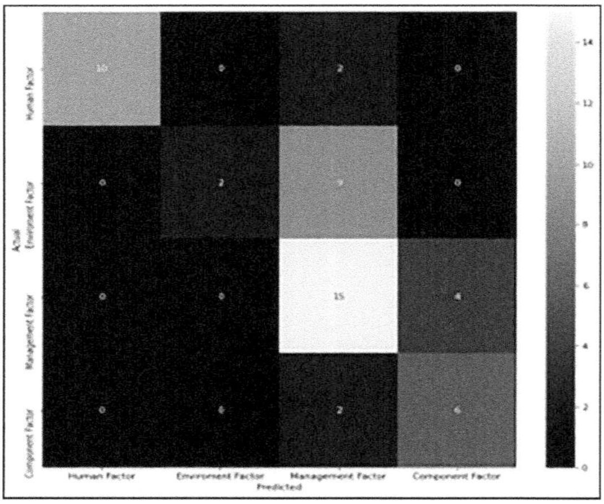

**FIGURE 11.6** (b) Confusion matrix.

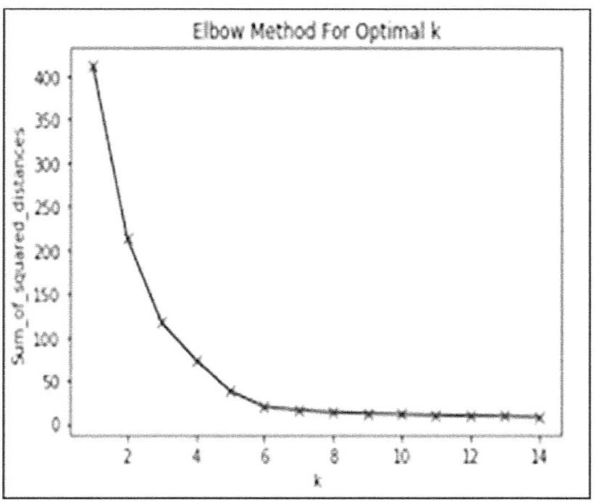


**FIGURE 11.6 (c)** Elbow plot.

Since it is a multiclass classification problem, linear SVC performs better. The accuracy from the linear SVC model we got is 66%. From the elbow method graph, we concluded the number of clusters we should make.

### *11.5.3 BOW-TIE ANALYSIS*

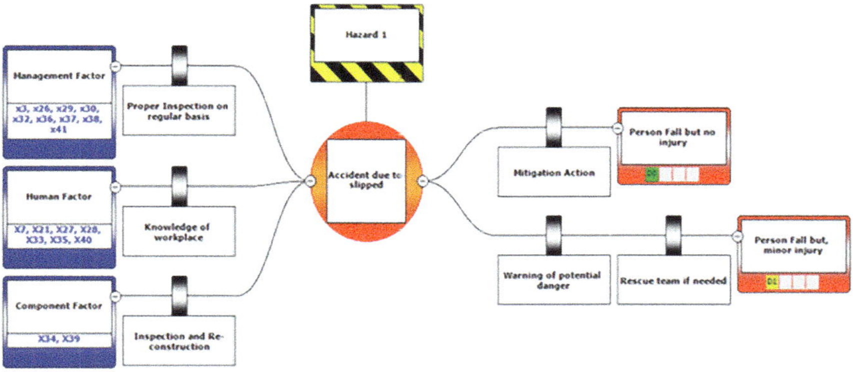


**FIGURE 11.7 (a)** Bow-tie diagram for slipped event.

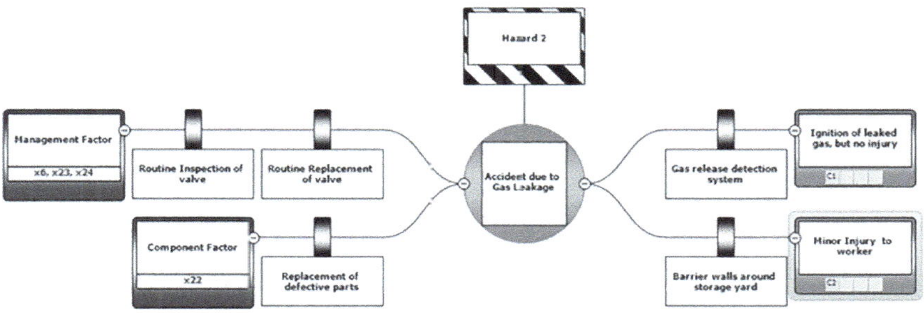


**FIGURE 11.7 (b)** Bow-tie diagram for leaked gas event.

We have used BowTieXP software to draw bow-tie diagrams for each top event. We observed that most of the accidents happened due to management errors. And the top events include (slipped, fell, fall, broken, damage, fire, breakage). Figure 11.7a and b are bow-tie diagrams for a slipped event and gas leakage, respectively, that lead to an accident. We suggested protective barriers for each factor on the left of bow-tie and protective barrier in case an accident already happened. Also, we categorize the root causation into four levels of factors shown on the left of the bow-tie diagram.

## 11.6 DISCUSSION

We collected more than 500 accident text data from the workplace at a distinct working site. Data comprise raw text data, so we first did preprocess like stop word removal, convert to lowercase and punctuation removal, tokenization, stemming, and lemmatization, thus bringing the data in a correct usable format. In order to find out the root causation and object involved in an accident, we applied a custom-trained NER model on annotated 450 text data with the label of root cause and object. This custom-trained NER model gave $F_1$ score of 0.84 which is better than compared with a pretrained NER model with $F_1$ score of only 0.51. Also, we find out the potential sentence which contains both root cause word and object using regular expression technique. To categorize the text containing causation words, we first applied the elbow method to get the optimal number of clusters, we got the output of 4–6. Linear SVC gives good accuracy over NB and RF classifiers. The accuracy we got from linear SVC is 66%. Finally, we build a bow-tie diagram using open-source BowtieXP software for better visualization of accident reports. Also, we find the most frequently occurring words in our whole corpus, called top event related to accidents like fire, fall down, slipped, and gas leakage. On the left of top event, we suggested preventive measures for the threat, and on the right, we suggested mitigative measures so that we can prevent future danger or injury.

## 11.7 CONCLUSION

In a workplace, workers experience varying levels of stress. As we know accidents nowadays happen quite often at every worksite. In this study, we used an integrated approach of NLP, TF–IDF vectorizer linear SVC, and qualitative bow-tie analysis. Our approach is different from other basic methods of quantitative risk analysis like Part of Speech tagging, predefined named entity model that is also capable of Information extraction from text. We applied a custom-trained NER model after annotating over 450 text data, thus got a better $F_1$ score over the predefined NER model. For classification, we used TF–IDF vectorizer and linear SVC model. The accuracy of linear SVC was better than other models like NB and RF classifier. Also, the accuracy will be further increased when we consider more elements of events that cause accidents such as weather and environmental factors. In order to visualize these accidents/incidents, root causes, and their categories, bow-tie methodology is applied. In the future, the proposed method can be employed in different sectors such as healthcare risk analysis and proper safety management in organizations.

## ACKNOWLEDGMENTS

We acknowledge the Centre of Excellence in Safety Engineering and Analytics (CoE-SEA) (www.iitkgp.ac.in/department/SE), IIT Kharagpur, and Safety Analytics and Virtual Reality (SAVR) Laboratory (www.savr.iitkgp.ac.in) of Department of Industrial and Systems Engineering, IIT Kharagpur, for experimental/computational and research facilities for this work. We would like to thank the management of the plant for providing relevant data and for their support and cooperation during the study.

## KEYWORDS

- **accident/incident reports**
- **root causes**
- **risk analysis**
- **natural language processing**
- **classification**
- **bow-tie analysis**

## REFERENCES

1. Abdi Z, Ravaghi H, Abbasi M, Delgoshaei B, Esfandiari S. Application of bow-tie methodology to improve patient safety. *International Journal of Health Care Quality Assurance* 2016;29(4):425–440.
2. Bertke SJ, Meyers AR, Wurzelbacher SJ, Measure A, Lampl MP, Robins D. Comparison of methods for auto-coding causation of injury narratives. *Accident Analysis and Prevention* 2016;88:117–123.

3. Chokor A, Naganathan H, Chong WK, Asmar, ME. Analyzing Arizona OSHA injury reports using unsupervised machine learning. *Procedia Engineering* 2016; 145:1588–1593.
4. Fan H, Li H. Retrieving similar cases for alternative dispute resolution in construction accidents using text mining techniques. *Automation in Construction* 2013;34:85–91.
5. Goh YM, Ubeynarayana CU. Construction accident narrative classification: An evaluation of text mining techniques. *Accident Analysis and Prevention* 2017;108(July):122–130. Retrieved from http://dx.doi.org/10.1016/j.aap.2017.08.026
6. Haslam RA, Hide SA, Gibb AGF, Gyi DE, Pavitt T, Atkinson S, Duff AR. Contributing factors in construction accidents. In *Applied Ergonomics*. 2005, Vol. 36, pp. 401–415.
7. Hickman L, Thapa S, Tay L, Cao M, Srinivasan P. Text preprocessing for text mining in organizational research: Review and recommendations. *Organizational Research Methods* 2020;25;114–146. https://doi.org/10.1177/1094428120971683
8. Kim S, Bracewell RH, Wallace KM. A framework for automatic causality extraction using semantic similarity. In: *2007 Proceedings of the ASME International Design Engineering Technical Conferences and Computers and Information in Engineering Conference, DETC2007*. 2008, Vol. 2 Part B, pp. 831–840.
9. Kim S, Lee J, Kang C. Analysis of industrial accidents causing through jamming or crushing accidental deaths in the manufacturing industry in South Korea: Focus on non-routine work on machinery. *Safety Science* 2021;133(October 2020):104998.
10. Niu S. Ergonomics and occupational safety and health: An ILO perspective. *Applied Ergonomics* 2010;41(6):744–753.
11. Singh V, Verma A. A review, simple meta-analysis and future directions of safety climate research in manufacturing organizations. *International Journal of Occupational Safety and Ergonomics* 2020:26(4):678–704.
12. Taylor JA, Lacovara AV, Smith GS, Pandian R, Lehto M. Near-miss narratives from the fire service: A Bayesian analysis. *Accident Analysis and Prevention* 2014;62:119–129.
13. Tixier AJP, Hallowell MR, Rajagopalan B, Bowman D. Application of machine learning to construction injury prediction. *Automation in Construction* 2016;69:102–114.
14. Tixier AJP, Hallowell MR, Rajagopalan B, Bowman D. Automated content analysis for construction safety: A natural language processing system to extract precursors and outcomes from unstructured injury reports. *Automation in Construction* 2016;62(2016):45–56.
15. Yang L, Li K. Safety risk analysis of railway accident with text-based bow-tie model. In: *Proceedings of 2020 IEEE 3rd International Conference of Safe Production and Informatization, IICSPI 2020*, 2020, pp. 200–204. https://doi.org/10.1109/IICSPI51290.2020.9332329
16. Zou Y, Kiviniemi A, Jones SW. Retrieving similar cases for construction project risk management using Natural Language Processing techniques. *Automation in Construction* 2017;80(September 2016):66–76.

## CHAPTER 12

# Finger Knuckle Print of an Emerging Person Recognition Trait for Online Applications

BRAJESH KUMAR SINGH[1*], ANIL KUMAR[2], and SUDESHNA CHAKRABORTY[2]

[1]*Department of Electronics and Communication Engineering, Galgotias college of Engineering and Technology, Greater Noida, Uttar Pradesh*

[2]*Department of Computer Science and Engineering, Galgotias University, Greater Noida, Uttar Pradesh, India*

*Corresponding author

## ABSTRACT

The human being has several biometric traits that can be used to identify a person such as fingerprint, finger vein, hand vein, hand geometry, face, iris, finger knuckle print, palm print, ear, gait, retina, and so on but the hand-based biometric system is highly acceptable due to easy acquisition, ease-to-use, and leading recognition performance. The fingerprint-based biometric system is an older one and is often used type for person recognition/identification. Apart from various advantages, the fingerprint trait has three major drawbacks. First, the fingerprint ridges are rugged in the hard-working people like cultivator, industrial labor, machine mechanics, rickshaw pooler, and so on. Second, the skin elasticity is not constant throughout the seasons; and third, applied thumb pressure is not always the same, cause dispersion of thumb skin is varied, resulting the extracted feature sets are also varied. Hence, intraclass variation occurs in the matching score and system performance degrades. The biometric domain

Machine Learning Applications for Data Analysis in Healthcare Systems: Sudeshna Chakraborty, Jyotsna Singh, Praveen Kumar Shukla, Prasenjit Chatterjee (Eds.)


DOI: 10.1201/9781779643193-12

researchers consider the above issues about the fingerprint biometric and nowadays person identification using finger knuckle print biometric is very popular among the biometric research community. This chapter presents an overview of this emerging biometric and available recognition methodology to identify people.

## 12.1 INTRODUCTION

The finger knuckle print (FKP) image is an inherent skin pattern available at the backside ofthe finger joint, recently this pattern has been quite popular and attracts the biometric research community due to its high acceptability and reliability [1]. The popularity gained by FKP is not only the cause of acceptability and ease of use, apart from that, it is not rugged like a fingerprint and easily can be captured using a normal resolution imaging device [2].

Woodard and Flynn [3] are the first scholars to make use of the finger knuckle surface to recognize a person. They used the Minolta 900/910 sensor to create a 3D finger back database. This sensor simultaneously captures and register image into size 640 × 480 and 640 × 480 of 24-bit color intensity, respectively. The sensor measures 213 mm × 413 mm × 270 mm and weighs around 11 kg. Due to its large size and weight, it is not suitable for commercial use of the biometric system. At the time of data collection, the sensor is placed on a flat wall at positioned approximately 1.3 m and is covered with a black cloth. Before the data collection was conducted, the subject was asked to remove all jewelry to ensure that the sensor does notscatter when in contact with the reflective surface of the jewelry. During the experiment, thesubjects were asked to place their right hand against the wall and keep it there as the image is captured. After that, they remove their hand from the wall and return it to the same position. The researchers collected a total of 1191 hand range images which are available publicly. In this image one is a taken a sample of hand color images of size 640 × 480 pixels. Second imageis taken as a pseudo intensity of the same hand image of size 640 × 480 using a polygonal meshand third image is a surface detail detected near a knuckle.

The only requirement for finger placement is such that there is space between two adjacent fingers. No constraints were used for capturing the images. They used a 3D range image of the finger to get the curvature surface representation. The normalized correlation coefficient was then used to compare the similarity of the images.

In 2009, Kumar and Ravikanth [4] developed a system using a digital camera to acquire finger-back surface images. The imaging system is focused against a white background under uniform illumination and the camera was fixed at a 20-cm distance from the imaging surface. The nonuniform illumination of objects at the hand boundaries reduces the performance of the acquired image. This is achieved by using a fixed light source above the hand. The resolution was 1600 × 200 pixels of the acquired FKP image. The subject is asked to place their hand on the support with the backhand facing the sensor.

There is a six-finger geometry and each has four fingers, resulting in 24 fingers being used to extract features. The features include finger length, widths, perimeter, and area along with the knuckle feature. The knuckle feature is extracted by applying the subspace analysis method onthe finger back surface and using these features for person classification. Although these methods were able to identify the unique features of the outer finger surface, they were not ableto provide a practical solution for implementing an efficient system.

The proposed method [3] focuses on the 3D shape information of FKP and not fully on the texture information while subspace analysis methods used in Ref. [4] may not be able to extract efficient features of FKP.

## 12.2 FKP RECOGNITION ALGORITHMS

The state-of-the-art method of FKP recognition algorithms may be classified into four categories, namely, (1) subspace-based methods, (2) coding methods, (3) other methods, and (4) recent trend method. Some of the algorithms are discussed as follows.

### *12.2.1 SUBSPACE-BASED FKP RECOGNITION ALGORITHM*

The subspace-based technique used to transform a raw image is into a low-dimensional subspace. The resultant subspace is used as a feature vector for classification by learning and training the data. These algorithms extract global information from an entire image using each subspace. These techniques are expected to enable the development of region-based identification methods. This technique uses principal component analysis (PCA), linear discriminant analysis (LDA), and independent component analysis (ICA) to extract subspace coefficient features. It is used to represent feature vectors, feature matching distancemeasures or classifiers, and also for dimension reduction of the extracted feature set.

A novel method called weighted linear embedding (WLE) was proposed [5] to recognize FKP.This method combines the Fisher criterion with the manifold learning criterion to extract features from a linear model. The concept of the manifold learning theory states that the importance of local information is more significant than nonlocal information. This concept is implemented and both local and nonlocal features are extracted from the data sets. The author uses Gaussian weighing to mapping procedure that combines nonlocal and local information. It is commonly used to find a mapping vector that is suitable for estimating the weighted class scatters maximized. The classification is done on the base of the nearest neighbor classifier. The authors have also tested the proposed algorithm on palmprint and have made a comparative study of PCA, LDA, LDE, and WLE. The WLE is tested on the right index finger of 1000 persons and achieved a recognition rate of 78.2%.

In Ref. [6], the author proposed a new FKP recognition work in which the author uses wavelets for image analysis and pattern recognition and also uses PCA to reduce the Gabor feature into low-dimensional space. Further, the author applies orthogonal linear discriminant analysis (OLDA) transformation in PCA subspace to classify the image using nearest neighbor and achieve 98% accuracy. This chapter also compares the individual finger performance and shows that the left index finger has better performance than the right index finger.

An author presented a method that simultaneously measures distances and angles between theimage data vectors to study the similarities between them [7]. An image angle measurement method that is focused on a shifted image space is proposed to highlight the distinction betweenthe angles of different data. Both the distance and angle are obtained through a parallel fusionprocedure that uses the complex locality-preserving projections to extract low-dimensional features. Orthogonal complex locality preservation projections or OCLPP is a subspace method that can minimize the redundant information in each image and four images were randomly selected for the recognition process and achieved a recognition rate up to 88%. Theauthor also compares the proposed work with the subspace method like PCA, CPCA, CLPP, and OCLPP. A summary of subspace-based technique is given in Table 12.1.

**TABLE 12.1** Summary of Subspace-Based Technique

| S. N. | Author | Feature Extraction | Evaluation Parameter | Results |
|---|---|---|---|---|
| 1 | Yin et al. [5] | WLE | KNN classifier | Recognition rate 78.2% |
| 2 | Wankou et al. [6] | Gabor and OLDA | KNN classifier | Recognition rate 98.06% |
| 3 | Jing et al. [7] | Orthogonal complex locality-preserving projections | Fused angle and Euclidean distance | Recognition rate 87.87% |

### *12.2.2 CODING-BASED FKP RECOGNITION ALGORITHM*

In the coding-based technique, the FKP image is segmented into several subimages, and information like orientation or edges is extracted from each subdivided image. Several codingalgorithms are proposed in the literature to improve the recognition performance of FKPs. These coding techniques are widely used for palmprint recognition and found good recognition results. The finger knuckle surface is very rich in lines and creases and highlyunique in individuals. These techniques of coding are mainly based on the concept of iris code.

A research work exploited the local feature in comparison to the global feature for reliable performance [8]. The preprocessing step helps to minimize the effects of light variation on the texture features and improves the overall performance. To avoid wrapping around an inherent modulo operation, finite radon transform and modified finite radon transform (MFRT) are commonly used to determine the orientation of the grease lines or knuckle lines ina local neighborhood region. The dominant direction at each pixel is computed and coded intobinary bits called knuckle code. The normalized Hamming distance is used to compute the similarity measurement between two FKP and these techniques achieved 98.6% accuracy.

A method [9] is proposed to capture an FKP image and automatically align the FKP image with a local coordinate system. The bottom of the FKP image is stable because of the acquisition method. Hence, this is considered the *x*-axis of the ROI coordinate system by fitting this boundary as a straight line. A model for the convexity of FKP was established. The resulting magnitude will reach the minimum at the center of the phalangeal joint and it can be used toset the coordinate system's *y*-axis. After the coordinate system is fixed, an ROI subimage of size 110 × 220 is cropped. Then orientation information is extracted using Gabor filtering andrepresented the information in the form of competitive code. The angular distance is used for matching between two FKP and finally an energy efficiency ratio (EER) of 1.09% was achieved.

An improved competitive and magnitude code is developed in 2010 [2]. This uses Gabor filters to extract the magnitude and orientation information. This information is further used toset up a code map that is based on the competitive code. This computed angular and magnitudedistance is fused and the minimum distance is considered the final distance. This proposed algorithm is compared with one standard coding algorithm and found better performance.

The concept of the monogenic code was proposed [8] to fast features extract from a two-dimensional signal $f(x,y)$. The monogenic signal is defined as the combination of $f(x,y)$ and its Riesz transform which is a vector-valued extension of the Hilbert transform in the 2D Euclidean space. The code represents the 3 bits of the monogenic signal that were extracted from the pixel. It shows the phase information and orientation of the pixel. This algorithm alsoshows similar accuracy in comparison to the state-of-the-art method of FKP recognition.

In Ref. [10], the author proposed a new coding method that has the advantages of high accuracy, robustness, and compactness. To find the above code, Riesz transform identifies andcharacterizes the visual patterns of the FKP image and also encodes the local patches using second-order Riesz transform. A 6-bit coding scheme namely the Riesz Compcode was developed. This Riesz Compcode is a 6-bit coding scheme that combines the advantages of Riesz transformation and Compcode in characterizing the local FKP feature together. A normalized Hamming distance is computed at the matching stage of the proposed work and this scheme of the coding algorithm shows better performance in terms of recognition accuracy when compared with another coding algorithm.

Nigam et al. [11] used pixel-based feature information around a specified area of the local blob, edge, or corners for FKP recognition. This technique of FKP recognition is sensitive to illumination but unguarded to rotation. It proposed a new feature extraction technique using vertical code and horizontal code. The proposed work extracts feature by taking derivatives along the *x*-axis and *y*-axis, respectively, and are represented by *vcode* and *hcode*. The extracted *vcode* and *hcode* are used by gradient ordinal relation pattern (GORP) and star gradient ordinal relation pattern (SGORP) to compute four types of features and are represented by $vcode_{GORP}$, $hcode_{GORP}$, $vcode_{SGORP}$ and $vcode_{SGORP}$. These four extracted features form a feature vector which is used for matching. Zhang et al. [9] used various coding techniques for FKP authentication and tested on a big dataset, and finally achieved a leading authentication performance. A summary of coding-based technique is given in Table 12.2.

**TABLE 12.2** Summary of Coding-Based Technique

| S. N. | Author | Feature Extraction | Evaluation Parameter | Results |
|---|---|---|---|---|
| 1 | Zhang et al. [2] | MFRT | Normalize hamming distance | Recognition rate 98.6% |
| 2 | Lin et al. [8] | Competitive code based on Gabor filter | Angular distance | EER 1.09% |
| 3 | Lin et al. [9] | Improved competitive code and magnitude code | Angular distance | EER 1.47% |
| 4 | Lin et al. [8] | Monogenic code | Hamming distance | EER 1.72% |
| 5 | Lin et al. [10] | Riesz transform-based coding | Hamming distance | EER 1.26% |

### *12.2.3 OTHER RECOGNITION ALGORITHM*

In the literature, several image processing techniques are used to extract texture features, local, global, and line features from FKP to either independently or combined manner. The local and global combined information provides more information and hence better recognition performance.

A technique known as local global information combination (LGIC) was proposed [12], and based on psychophysics and neurophysiology study, the author states that both local and global information are crucial for FKP recognition. The local feature is orientation information whichis extracted by the Gabor filter using four scales and six orientations wavelet. This procedure is suitable for extracting features from images with an abundant line-like structure available and has advantages like high accuracy, robustness to light intensity variation, and computation efficiency. The global feature is extracted by increasing the scale to infinity of the Gabor filter and obtaining the Fourier coefficient from the transformed FKP image.

The matching strategy uses two code maps, one is the angular distance which is computed based on the normalized Hamming distance, and the other is band-limited phase-only correlation (BLPOC) which is used to measure the similarity between Fourier transforms (global information) of the images.Thus, two separately calculated matching distances $d_1$ and $d_2$ from local and global features combine according to the matching weight rule and this algorithm achieved EER as low as 0.402%.

In Ref. [13], the author tried to extract the orientation feature from the FKP image using a Gabor filter bank having five different scales and eight different orientations and keeping all other parameters constant. Further, the author applied the PCA for feature dimension reduction. Since a combination of PCA and LDA provides a good effecton feature selection LDA is applied to PCA weights. The Euclidean distance is used for matching. The proposed method of FKP recognition is tested over all four fingers image and itis observed that the right middle finger has a better recognition rate of 75.25%. A different combination of feature-level fusion is carried out for all fingers and the maximum recognitionrate was 98.79% achieved.

A novel FKP recognition technology proposed robust FKP feature extraction and matching algorithm based on speeded-up robust features (SURF) [14]. This novel feature is extracted by following two steps: first, a coordinate system is defined based on the local convex direction map of FKP to align the FKP image and a region of interest is cropped. Second, the key points are detected using a fast Hessian detector to which orientation was assigned accordingly to the Haar wavelet responses inside a neighbor circle area of a detected key point, and an orientation invariant descriptor is constructed for each key point. In this matching method, the distance of the second-closest neighbor is matched with the distance of the first-closest neighbor. All matches are performed in which the distance ratio between the two is less than 0.6. Thus, the tentative correspondence between two key point sets of training image and template image is obtained. The RANdom SAmple Consensus (RANSAC) is formulated to remove the false matching and the final matched pair is used to determine the consistency of the palm images. This method is invariant to rotate and scale transform and provides an accurate and robusticity of 90.63% for validation and 96.91% for identification purposes.

A novel method was proposed to identify individuals using a single FKP [15]. A major advantage of this method is to reduce the huge database size and this proposed method trains the classifier by describing the edges of an image with discontinuities in the grey levels. A self-defined convolution template of size 3 × 5 as a gradient operator is used in the spatial domain to minimize the noise and extract the lines feature. The method was then used to prevent making a wrong decision due to the varying position and different standing posture of the acquisition device during the collection of images. Eight different images were obtained by translation operation and one image original, as a whole nine image is used for confirming the user's identity using the principle of the maximum cross-correlation coefficient. The experiments conducted on the FKP obtained a recognition rate of 95.68% at a 30-threshold value, which revealed that it is a reliable and accurate biometric trait.

Morales et al. [16] presented a novel approach for FKP authentication. In this technique, the author uses a Gabor transform to enhance the orientation information of the image, and SIFT descriptor is used to extract and represent the feature sets. El-Tarhouns et al. [17] applied multiscale LBP and PCA techniques to extract the local and global features from the FKP image and a matching attempt can be considered as a combination of both global and local features.

In Ref. [18], the author proposed a multimodal biometric system using FKP and Palmprint biometric trait. This model extracted various features from both and computed scores are combined using score level fusion. Various combinations of experiments are conducted and found that SURF plus bidirectional empirical mode decomposition feature extraction technique provides better performance. A summary of other techniques for FKP recognition is given in Table 12.3.

**TABLE 12.3** Summary of Other Techniques to FKP Recognition

| S. N. | Authors | Feature Extraction | Evaluation Parameter | Results |
|---|---|---|---|---|
| 9 | Le-Qing et al. [19] | Speeded-up robust features (SURF) | RANSAC matching strategy | Recognition rate 96.91% |
| 10 | Zhang et al. [12] | LGIC technique | Hamming distance and BLPOC | Recognition rate 99.31% |
| 11 | Zhao et al. [20] | Edge detection | Cross-correlation coefficient | Recognition rate 95.68% |
| 12 | Shariatmadar et al. [9] | Gabor filter + PCA + LDA | Euclidean distance | Recognition rate 89.90% |

## 12.3 BIOMETRIC SYSTEM BASIC BLOCK DIAGRAM

Originally, the biometric system is a pattern recognition system that obtains the biometric datafrom an individual, draws out a primary characteristic set from the data, differentiates this characteristic set against the characteristic set(s) reserved in the database, and results in a decision based on the differentiation.

A biometric system thus has the following basic modules: a raw data acquisition module, a quality improvement module (preprocessing module), a matching module, feature extraction module, and a database module. Figure 12.1 describes the basic block diagram of the biometric system.

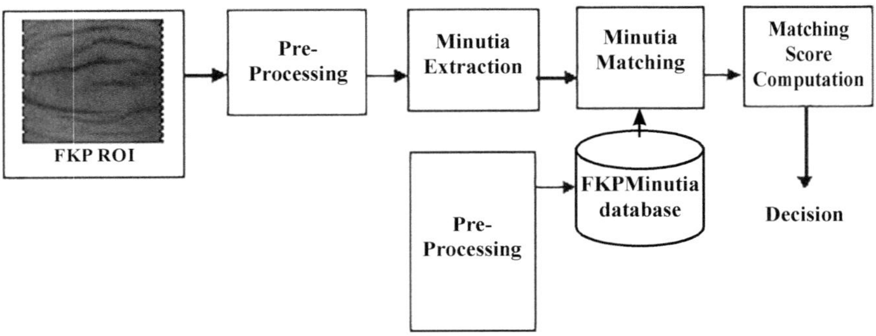


**FIGURE 12.1** Basic block diagram of biometric system.

### 12.3.1 DATA ACQUISITION

Raw data from the individual may be obtained by using an appropriate sensor, a camera,or a scanner. This module must be capable of acquiring good-quality images for the biometric system to provide good performance.

### 12.3.2 PREPROCESSING

The function of this module is to assess the quality of the raw data obtained from the previous stage. Any preprocessing algorithm may be applied to the raw data to enhance itsquality so that it is suitable for feature extraction.

### 12.3.3 FEATURE EXTRACTION

Next, a suitable algorithm is applied to the preprocessed image to extract the features (a compact representation of the preprocessed image) that can be used to uniquely identify a person. The extracted features are stored as templates in the database.

### *12.3.4 MATCHING*

For a given test or a query image, the features are extracted and are compared with the reserved templates of the database to measure the level of closeness between the template and the extracted features of the clamming person. A decision is then made as to whether the individual or the user is a genuine (authorized or enrolled user) or an impostor(unauthorized user) depending on the observed closeness measure.

### *12.3.5 DATABASE*

The feature set extracted from the raw biometric data (i.e., the template) is stored in the database along with some additional information (such as name, Personal Identification Number, address, etc.) representing the user.

## 12.4 HARDWARE IMPLEMENTATION AS IOT DEVICE

Raspberry Pi is a small card size computer that operates on Debian Linux. It is easy to use andlearn. The operating system of Raspberry Pi can be installed easily by just adding a simple SDcard. The SD card will contain the OS, and it will be added into the Pi board and to power this board a simple USB mobile charger will be required, after this just connect the keyboard and mouse and a TV or computer will be connected by an HDMI cable and it is ready for use. If we need to work on a Windows platform, we might need Putty and VNC Viewer application toopen the Pi desktop on the Windows platform. Now, as the Pi desktop is opened it can do multiple things you can send emails or create spreadsheets. Raspberry Pi provides us with an opportunity to code with scratch or Python. As we are here working with an optical fingerprintsensor R307, we need to import multiple libraries like OpenCV or others. Raspberry Pi contains a command prompt from which we can import the libraries. It contains Ides so that we can code easily and run to see the function happening properly. So here we can start our work; the fingerprint sensor will capture the image of the fingerprint and store it in the PI and with the help of OpenCV which will fetch the image and from where we can apply the image preprocessing like binarization and thinning and extract features and then we can matchit with the already stored features.

## 12.5 CONCLUSIONS

In the literature, several algorithms are also available to recognize FKP biometric trait and recognition potential of this trait are very competitive as compared to other biometric traits and it is not rugged like fingerprint biometric due available at the outer surfaceof finger. This biometric trait can be used to design a biometric system design for online personverification as an Internet of Things device.

## KEYWORDS

- **finger knuckle print**
- **person identification**
- **biometric system**
- **subspace method**
- **coding method**

## REFERENCES

1. Zhang D, Guo Z, Lu G, Zhang L, Zuo W. An online system of multispectral palmprint verification. *IEEE Transactions on Instrumentation and Measurement* 2009;59(2):480–490.
2. Zhang L, Zhang L, Zhang D, Zhu H. Online finger-knuckle-print verification for personal authentication. *Pattern Recognition* 2010;43(7):2560–2571.
3. Woodard DL, Flynn PJ. Finger surface as a biometric identifier. *Computer Vision and Image Understanding* 2005;100(3):357–384.
4. Kumar A, Ch R. Personal authentication using finger knuckle surface. *IEEE Transactions on Information Forensics and Security* 2009;4(1):98–110.
5. Yin J, Zhou J, Zhong J, Yang J. Weighted linear embedding and its applications to finger-knuckle-print and palmprint recognition. In: *2010 International Workshop on Emerging Techniques and Challenges for Hand-Based Biometrics*. IEEE, 2010, pp. 1–4.

6. Yang W, Sun C, Wang Z. Finger-knuckle-print recognition using Gabor feature and MMDA. *Frontiers of Electrical and Electronic Engineering in China* 2011;6(2):374–380.
7. Jing X, Li W, Lan C, Yao Y, Xi C, Lu H. Orthogonal complex locality preserving projections based on image space metric for finger-knuckle-print recognition. In: *2011 International Conference on Hand-Based Biometrics*. IEEE, 2011, pp. 1–6.
8. Zhang L, Li H, Shen Y. A novel Riesz transforms based coding scheme for finger-knuckle-print recognition. In: *2011 International Conference on Hand-based Biometrics*, IEEE, 2011, pp. 1–6.
9. Zhang L, Zhang L, Zhang D. Finger-knuckle-print: A new biometric identifier. In: *2009 16th IEEE International Conference on Image Processing (ICIP)*. IEEE, 2009, pp. 1981–1984.
10. Zhang L, Zhang L, Zhang D. Monogeniccode: A novel fast feature coding algorithm with applications to finger-knuckle-print recognition. In: *2010 International Workshop on Emerging Techniques and Challenges for Hand-Based Biometrics*. IEEE, 2010, pp. 1–4.
11. Nigam A, Tiwari K, Gupta P. Multiple texture information fusion for finger-knuckle-print authentication system. *Neurocomputing* 2016;188:190–205.
12. Zhang L, Zhang L, Zhang D, Zhu H. Ensemble of local and global information for finger–knuckle-print recognition. *Pattern Recognition* 2011;44(9):1990–1998.
13. Shariatmadar ZS, Faez K. A novel approach for Finger-Knuckle-Print recognition based on Gabor feature fusion. In: *2011 4th International Congress on Image and Signal Processing*. IEEE, 2011, Vol. 3, pp. 1480–1484.
14. Michael GKO, Connie T, Jin ATB. Robust palm print and knuckle print recognition system using a contactless approach. In: *2010 5Th IEEE Conference on Industrial Electronics and Applications*. IEEE, 2010, pp. 323–329.
15. Zhang L, Zhang L, Zhang D. Finger-knuckle-print verification based on band-limited phase-only correlation. In: *International Conference on Computer Analysis of Images and Patterns*, Springer: Berlin, Heidelberg, 2009, pp. 141–148.
16. Morales A, Travieso CM, Ferrer MA, Alonso JB. Improved finger-knuckle-print authentication based on orientation enhancement. *Electronics Letters* 2011;47(6):380–381.
17. El-Tarhouni W, Shaikh MK, Boubchir L, Bouridane A. Multi-scale shift local binary pattern based-descriptor for finger-knuckle-print recognition. In: *2014 26th International Conference on Microelectronics (ICM)*. IEEE, 2014, pp. 184–187.
18. Perumal E, Ramachandran S. A multimodal biometric system based on palmprint and finger knuckle print recognition methods. *International Arab Journal of Information Technology (IAJIT)* 2015;12(2).
19. Zhu L-Q. Finger knuckle print recognition based on SURF algorithm. In *2011 Eighth International Conference on Fuzzy Systems and Knowledge Discovery (FSKD)*. IEEE, 2011, Vol. 3, pp. 1879–1883.
20. Zhao R, Li K, Liu M, Sun X. A novel approach of personal identification based on single knuckleprint image. In *2009 Asia-Pacific Conference on Information Processing*. IEEE, 2009, Vol. 2, pp. 218–221.

# CHAPTER 13

# Face Mask Detection Using Deep Learning and Image Processing Algorithms

PALLAVI GOEL[1*], PRAVEEN KUMAR SHUKLA[2], AJITESH GAUTAM[1], ASHISH YADAV[1], and SHANTANU SINGH[1]

[1]*Department of Computer Science & Engineering, Galgotias College of Engineering and Technology, Greater Noida, Uttar Pradesh, India*

[2]*Department of IOT & Intelligent Systems, Manipal University Jaipur, Jaipur, Rajasthan, India*

*Corresponding author

## ABSTRACT

In recent years, the coronavirus has emerged as a highly concerning health issue for human beings due to its infectious nature and potential lethality. The virus is transmitted through exhalation, which spreads viral particles to surfaces that can be breathed in by susceptible individuals, leading to infection. Thus, it is critical to prevent ourselves and others around us from getting infected by practicing social distancing, hand cleanliness, and wearing masks to cover face in public. The widespread adoption of masks in public places globally highlights their importance in mitigating the transmission of the virus. The ability to identify people wearing masks is becoming increasingly important in public places. This research suggests a novel method for identifying face masks by combining the ResNet-50 model with a bilinear interpolation, bilateral filter, and fuzzy *c*-means segmentation. By improving details and segmenting the face and background areas in the input picture, our technique seeks to increase the

Machine Learning Applications for Data Analysis in Healthcare Systems: Sudeshna Chakraborty, Jyotsna Singh, Praveen Kumar Shukla, Prasenjit Chatterjee (Eds.)


DOI: 10.1201/9781779643193-13

reliability of face mask identification. We tested our approach on a publicly accessible dataset and found it to be effective, with encouraging results and high accuracy. Our method has potential applications in scenarios where face mask detection is required, such as airports, hospitals, and public transportation.

## 13.1 INTRODUCTION

Because of the COVID-19 pestilence, people all over the world need to take steps to stop it from spreading. Face masks are among the most efficient ways to stop the virus from spreading. But it can be hard to make sure that everyone in public places wears a mask at all times. In this situation, making face mask detection systems that work automatically can be a key part of lowering the risk of getting sick.

In this chapter, we propose a way to solve this problem by making a face mask detection system using ResNet-50, fuzzy *c*-mean segmentation, and bilinear interpolation. The proposed system is meant to automatically tell if people in an image are wearing face masks or not. It could be added to existing surveillance systems to keep track of people in real time. Overall, our research demonstrates the potential of using deep learning techniques and image processing algorithms to automate the face mask detection procedure, offering a practical solution to help containing the spread of COVID-19.

### *13.1.1 DEEP LEARNING*

It is a specialized branch of machine learning. Deep learning trains multi-layered neural networks to do complex tasks. Deep learning algorithms use vast amounts of data to automatically learn characteristics and patterns for accurate predictions and classifications.

Deep learning has transformed computer vision, NLP, speech recognition, and robotics. Deep learning has enabled accurate, scalable models for image classification, object identification, segmentation, language translation, and speech synthesis.

Deep learning algorithms work well with big data because they can learn from plenty of data. Deep learning has also advanced drug development, genetics, and customized medicine by analyzing difficult datasets.

### 13.1.2 CNN (CONVOLUTIONAL NEURAL NETWORK)

Convolutional neural networks (CNNs) serve the purpose of image classification and recognition. CNNs automatically extract features from images, unlike traditional neural networks. CNNs use convolutional layers to apply learnable filters to input images. These filters identify edges and textures, and the convolutional layer output is routed via nonlinear activation functions to build a feature map. Pooling layers downsample the feature maps to minimize their spatial dimensions while retaining the most relevant characteristics. CNNs finish with fully linked layers to classify learned features. Backpropagation with stochastic gradient descent reduces a loss function by modifying filter weights and biases.

CNNs excel at picture classification, object recognition, and segmentation. Natural language processing, voice recognition, and other applications employ them.

## 13.2 RELATED WORK

Proposal for three datasets on masked faces: mask occlusion face detection dataset, real-world masked face recognition dataset (RMFRD), and synthetic masked face recognition dataset [1]. These datasets would be free for both academia and industry to use to make different applications for masked faces. RMFRD is the biggest real-world facemask dataset in the world right now [2]. Study and overview of image segmentation techniques in CNN, covering region-based, feature extraction, clustering, and loosely supervised learning, with a forecast of image segmentation development trend [3]. Implementation of a technique using computers for the detection and segmentation of brain tumors in magnetic resonance (MR) images, reducing analysis time, and displaying the tumor's exact position and shape [5]. The creation of a technically enabled drone for recognition of mask and social distancing surveillance, which uses a quicker R-CNN model to find people who do not have their face masked and send out notifications to keep people apart [7] from creating a neural network approach (region proposal network) that works with the detection network to share full-image convolutional features allows for practically cost-free area proposals and achieves the highest item detection performance on PASCAL VOC 2007 and 2012 [9] masked face detection using MobileNetV2 architecture with two-layer max pooling, average pooling, with high training, and validation accuracy [10]. Using the ResNet-50 architecture, as a superresolution and classifier framework, arbitrary trigonometric transform-based convolution with multilayer perceptron is employed for tumor identification and classification, obtaining 98.14% accuracy rate.

## 13.3 PROBLEM FORMULATION

### *13.3.1 PROBLEM STATEMENT*

The COVID-19 has made us rethink our daily routines. Face masks are a popular and effective way to reduce infection risk. Face mask detection devices are in demand for safety. This project extends points to build a profound learning-based face mask detector to prevent coronavirus spread utilizing image processing algorithms for improved optimization.

### *13.3.2 PROBLEM DESCRIPTION*

Due to its contagious and sometimes fatal characteristics, the coronavirus has become a major health concern in recent years. Exhalation transfers viral particles to surfaces that susceptible people breathe in, infecting them. We must avoid people, wash our hands, and use masks to avoid this situation. Because to its enormous population density, India is the most affected by public mask use.

A convolutional neural network can detect face masks in vast workplaces. Real-world data trains and live video streams validate the model. Testing hyperparameters, more people at varied distances, and frame positioning determines the model's precision. Disinfectant and hand washing are also advised.

## 13.4 SYSTEM DESIGN

### *13.4.1 HIGH-LEVEL SYSTEM ARCHITECTURE*

1. Data collection: Gather a database of pictures of individuals wearing and not wearing masks.

2. Data preprocessing: Do form three parts of the existing dataset, that is, training, validation, and testing and resizing the photos, supplement the data (rotate, flip, alter brightness/contrast), and also implement the preprocessing image algorithms.
3. Feature extraction: Extract features from the images using a pretrained ResNet-50 model.
4. Fine-tune the model: Adjust the model by adding a few more layers to distinguish between "with mask" and "without mask" image classifications.
5. Model evaluation: Assess the performance of the model on the validation set and make any required adjustments to the hyperparameters.
6. Deployment: Deploy the model as an application that can analyze live video or still photos from a camera and decide whether or not individuals in them are wearing masks, then respond appropriately (e.g., alerting a security guard, denying entry to a building, etc.).

### 13.4.2 SYSTEM ARCHITECTURE

The following three components make up the whole of the proposed system in its entirety (Figure 13.1).

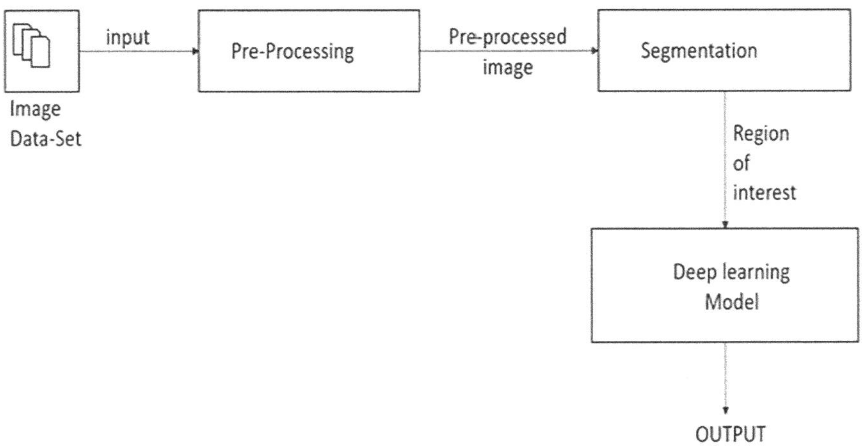


**FIGURE 13.1** System architecture for facemask detection.

### *13.4.3 DATASET*

For the purpose of this study, we are utilizing different datasets in order to produce results that are more precise and accurate. We are creating a more accurate dataset with the help of MaskedFace Net [12], face mask dataset [9], face mask detection data [10], face mask detection [11] 12k images dataset, and masked face recognition dataset and application. This dataset will have over 6000 images of people wearing masks and approximately 5100 images of people without masks. We planned the dataset in such a way that it would include both generic and medical masks, each of which would have a variety of colors and textures.

The dataset aims to detect whether people are wearing masks or not. Additionally, a mask-to-face deformable model is presented in this study, enabling the creation of additional masked face pictures using certain masks. This dataset provides a granularity of classification that permits mask-wearing analysis, making it a valuable resource for research on masked face recognition (Figure 13.2).

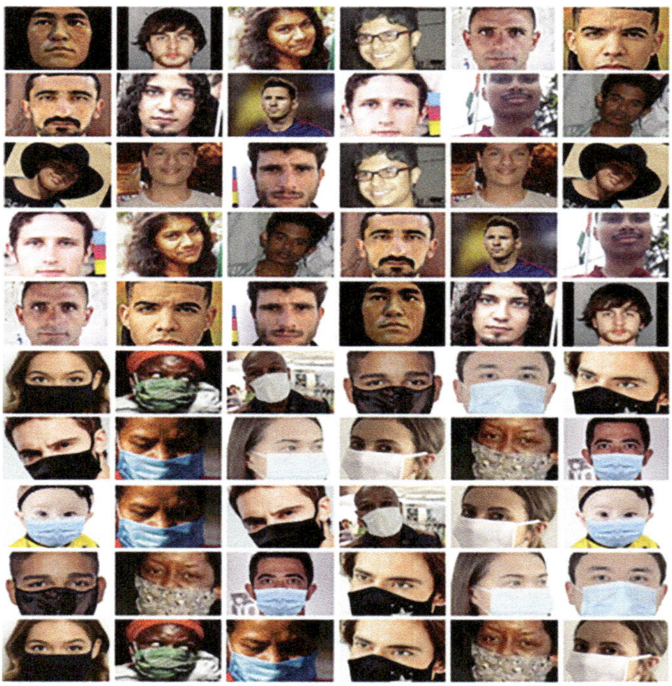

**FIGURE 13.2** View of the dataset.

## 13.4.4 PREPROCESSING AND SEGMENTATION

### 13.4.4.1 BILINEAR INTERPOLATION

Bilinear interpolation estimates pixel values between neighboring pixels on a regular grid in image processing and computer graphics. The weighted average of the four closest known pixels is used to estimate the value of an unknown pixel.

The method estimates the pixel value by weighting the distance between the unknown pixel and its nearby known pixels. The equation uses weights from each known pixel to the unknown pixel. The interpolated value takes into account the unknown pixel's proximity to a known pixel.

Bilinear interpolation helps resize photos. The technique estimates new pixel values between known pixels when a picture is enlarged, delivering a smoother result than duplicating pixels. Bilinear interpolation can estimate pixel values to preserve image quality when reducing image size (Figure 13.3).

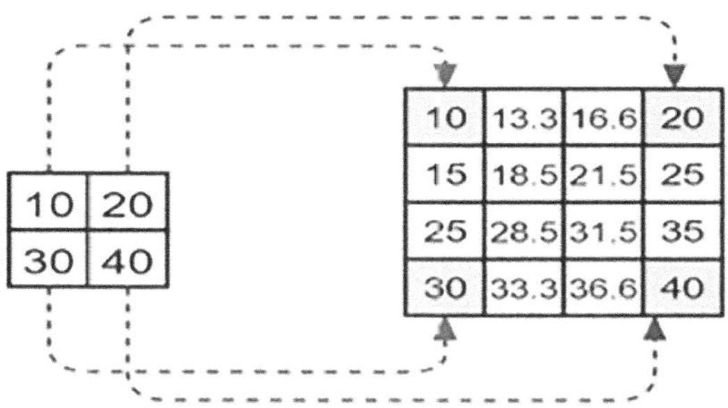


**FIGURE 13.3** Bilinear interpolation.

Overall, bilinear interpolation is a simple and effective method for estimating pixel values in image processing and computer graphics, particularly in situations where smooth transitions between adjacent pixels are desired.

### 13.4.4.2 BILATERAL FILTER

Image processing often uses the bilateral filter. It weights pixels based on their spatial and intensity distances and averages them. The filter smoothens photos while preserving borders, texture, and small details.

The bilateral filter calculates the space and intensity distance between the center pixel and the surrounding pixels. Bigger disparities in pixel values are given less weight when calculating intensity distance. Pixels closer together are weighted more in the Euclidean distance calculation. Multiplying the intensity and distance weights yields a weighted average of nearby pixel weighs smooth. Pixel outputs are weighted averages of surrounding pixels. Similar-intensity pixels nearby weighted more (Figure 13.4).

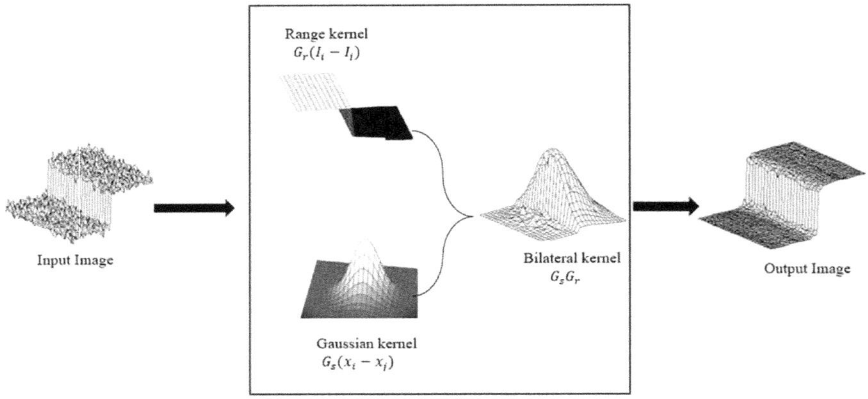


**FIGURE 13.4** Bilateral filter.

Overall, the bilateral filter is used in image processing applications such as denoising, texture removal, and picture enhancement. It smoothens pictures while keeping edges and other high-frequency details.

### *13.4.4.3 FUZZY C-MEAN SEGMENTATION*

Image segmentation often uses the unsupervised fuzzy *c*-mean segmentation (FCM) clustering algorithm. Data points can belong to several clusters with varying degrees of affiliation using the FCM method. FCM can partition images using pixel values. Each pixel's membership value reflects its cluster affiliation.

The algorithm updates the cluster centroids and membership values iteratively until convergence, having the aim of lowering the distance between the data points and the cluster centroids. FCM is capable of dealing with visual ambiguity and overlap, making it appropriate for picture segmentation.

It is extremely flexible as it allows the user to define the number of desired clusters and can be integrated with other techniques, such as geographical information, to improve segmentation (Figure 13.5).

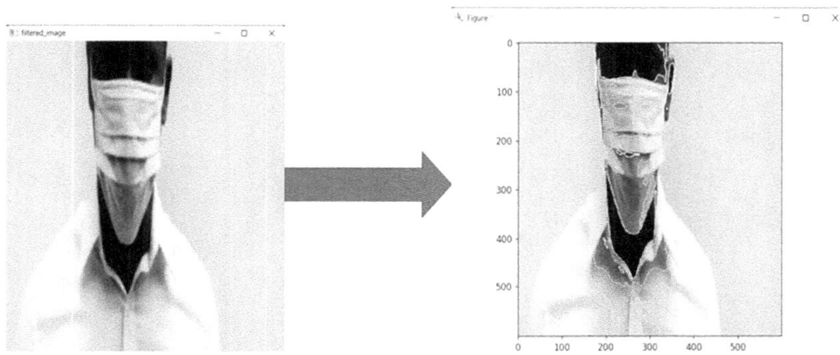


Filtered Image　　　　Segmented Image

**FIGURE 13.5** Fuzzy *c*-mean segmentation.

Overall, FCM is a powerful and widely used algorithm for image segmentation that offers flexibility and robustness in handling complex image data.

### *13.4.5 TRAINING MODEL*

This project was completed utilizing ResNet-50 as a training model. ResNet-50 is a deep neural network architecture that has garnered popularity in computer vision tasks since Microsoft Research Asia's introduction of it in 2015.

This 50-layer architecture is a variant of the ResNet family of neural networks, which is well-known for tackling the problem of vanishing gradients by introducing the skip connection model. By enabling information to flow from earlier to later network layers, these skip connections help preserve essential features and reduce the impact of vanishing gradients.

ResNet-50 has achieved cutting-edge performance on a variety of benchmark datasets, including ImageNet, COCO, and PASCAL VOC, making it a popular choice for image classification, object recognition, and segmentation applications. Furthermore, the ResNet-50 architecture is extensively used for transfer learning, which involves applying pretrained models to new datasets with less training data to train them.

In conclusion, ResNet-50 is an important advancement in the field of deep learning. It offers an efficient method for the training of very deep neural networks and accomplishes remarkable outcomes across a variety of different existing computer vision tasks (Figure 13.6).

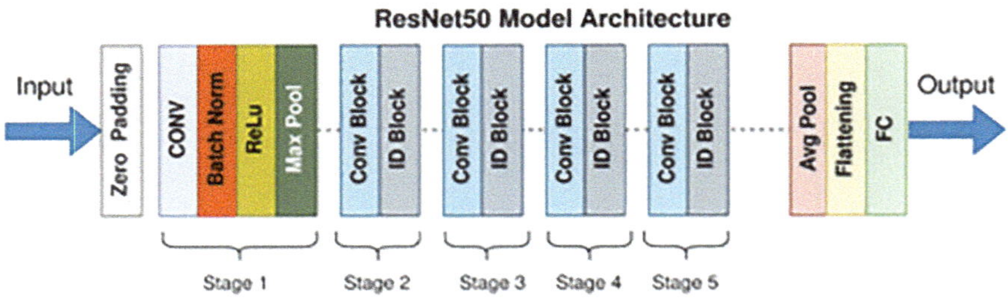


**FIGURE 13.6** Architecture of ResNet-50.

## 13.4 RESULT AND DISCUSSION

The training and validation loss and accuracy of every epoch is given below.

| Epoch | Training Loss | Training Accuracy | Validation Loss | Validation Accuracy |
|---|---|---|---|---|
| I | 1.3198 | 0.9680 | 1.3290 | 0.9688 |
| II | 0.2265 | 0.9946 | 0.1878 | 0.9939 |
| III | 0.0809 | 0.9975 | 1.1236 | 0.9827 |
| IV | 0.1360 | 0.9957 | 1.3337 | 0.9653 |
| V | 0.0672 | 0.9975 | 0.2728 | 0.9913 |
| VI | 0.0312 | 0.9984 | 0.6057 | 0.9896 |
| VII | 0.0334 | 0.9988 | 0.1432 | 0.9948 |
| VIII | 0.0169 | 0.9990 | 0.2503 | 0.9913 |
| IX | 0.0169 | 0.9992 | 0.2863 | 0.9939 |
| X | 0.0105 | 0.9994 | 0.2636 | 0.9887 |
| XI | 0.0467 | 0.9979 | 0.4122 | 0.9801 |
| XII | 0.0045 | 0.9994 | 0.5453 | 0.9896 |
| XIII | 0.1585 | 0.9840 | 0.2636 | 0.9853 |
| XIV | 0.0899 | 0.9963 | 0.1852 | 0.9939 |
| XV | 0.0621 | 0.9957 | 1.3052 | 0.9809 |
| XVI | 0.0898 | 0. 9973 | 0.7436 | 0.9740 |
| XVII | 0.0141 | 0.9988 | 0.5381 | 0.9887 |

After training the model, we tested it on a training dataset of 923 images and it detects people with mask and without mask with accuracy of 99.56 (Figure 13.7).

```
test_loss,test_acc=model.evaluate(test_generator,verbose=2)
print('\n test accuracy:',test_acc)

923/923 - 136s - loss: 0.1192 - accuracy: 0.9957 - 136s/epoch - 148ms/step

 test accuracy: 0.9956663250923157
```

**FIGURE 13.7** Accuracy analysis of model.

Our model is capable of detecting the people with and without mask with better accuracy (Figure 13.8).

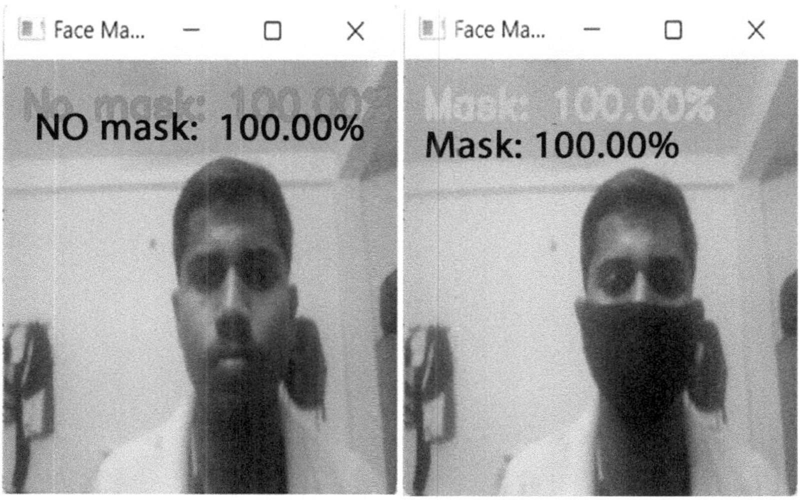


**FIGURE 13.8** Result.

## 13.5 CONCLUSION

This work concludes with a deep learning and image processing face mask detection method.

CNNs extract and classify features, while image processing detects masks. A massive dataset of pictures with and without masks tested the strategy.

The described method accurately detects face masks with few false positives. The device's real-time detection is ideal for airports, hospitals, and other mask-wearing environments.

This research increases computer vision and pattern recognition and combats infectious diseases like COVID-19 with an automatic and reliable face mask detection system. The strategy could increase public health and mask-wearing compliance, boosting safety and well-being.

Masks worn incorrectly or partially covered may cause false negatives with the suggested approach. Research could improve the algorithm's robustness. Using the suggested technology requires considering privacy, ethics, and human factors.

In conclusion, deep learning and image processing algorithms can detect face masks, and further research could affect public health and safety.

## KEYWORDS


- **COVID-19**
- **bilinear interpolation**
- **face mask bilateral**
- **filter**
- **fuzzy *c*-mean segmentation**
- **deep learning**
- **ResNet-50**


## REFERENCES


1. Wang Z, Wang G, Huang B, Xiong Z, Hong Q, Wu H, Yi P, et al. Masked face recognition dataset and application. arXiv preprint arXiv:2003.09093 (2020).
2. Yuheng S, Hao Y. Image segmentation algorithms overview. arXiv preprint arXiv:1707.02051 (2017).
3. Selvakumar J, Lakshmi A, Arivoli T. Brain tumor segmentation and its area calculation in brain MR images using K-mean clustering and Fuzzy C-mean algorithm. In: *IEEE-International Conference on Advances in Engineering, Science and Management (ICAESM-2012)*. IEEE, 2012, pp. 186–190.

4. Meivel S, Sindhwani N, Anand R, Pandey D, Alnuaim AA, Altheneyan AS, Jabarulla MY, Lelisho ME. Mask detection and social distance identification using internet of things and faster R-CNN algorithm. *Computational Intelligence and Neuroscience*. 2022;2022:2103975.
5. Ren S, et al. Faster R-CNN: Towards real-time object detection with region proposal networks. In: *Advances in Neural Information Processing Systems*. 2015, Vol. 28.
6. Shiekh B. Face mask dataset, version 1; September 2022. https://www.kaggle.com/datasets/ashishjangra27/face-mask-12k-images-dataset (accessed Mar. 22, 2023).
7. Chakraborty A.Face mask detection data, version 1; 2019. https://www.kaggle.com/datasets/aneerbanchakraborty/face-mask-detection-data/code (accessed Mar. 14, 2023).
8. https://github.com/cabani/MaskedFace-Net
9. Jangra A. Face mask detection ~12K images dataset, version 1; 2020. https://www.kaggle.com/datasets/shiekhburhan/face-mask-dataset (accessed March 20, 2023).

# Index

## C



## D